T0363371

DR JACKSON RYAN is a former molecular biologist and current science and tech journalist with a focus on longform and narrative non-fiction science writing. He served as the science editor at CNET.com between 2018 and 2023 and was the 2022 winner of the Eureka Prize for Science Journalism. His longform writing has appeared in the *New York Times*, the *Guardian*, *The Monthly* and *Nature*. He lives with a collection of more than 70 Christmas sweaters and zero dachshunds, the latter of which he hopes to rectify one day.

CARL SMITH is a science journalist working for ABC's Science Unit and Radio National. He makes audio documentaries, podcasts and written features with original photography. He's won a Young Walkley Award for Longform Journalism and jointly won a Eureka Prize for Science Journalism. Carl's been an ABC News cadet, a geneticist, a reporter on *Behind the News*, a 'journalist in residence' in Germany and an animated presenter on the ABC Education series *Minibeast Heroes*. His most recent audio series are *Strange Frontiers* and *Pacific Scientific*. He also co-writes and co-hosts the ethics podcast for kids, *Short & Curly*.

THE BEST AUSTRALIAN SCIENCE WRITING 2024

EDITORS
JACKSON RYAN
AND CARL SMITH

FOREWORD
COREY TUTT OAM

NEWSOUTH

UNSW Press acknowledges the Bedegal people, the Traditional Owners of the unceded territory on which the Randwick and Kensington campuses of UNSW are situated, and recognises the continuing connection to Country and culture. We pay our respects to Bedegal Elders past and present.

A NewSouth book

Published by
NewSouth Publishing
University of New South Wales Press Ltd
University of New South Wales
Sydney NSW 2052
AUSTRALIA
https://unsw.press/

Introduction © Dr Jackson Ryan and Carl Smith 2024
First published 2024
10 9 8 7 6 5 4 3 2 1

 A catalogue record for this book is available from the National Library of Australia

ISBN   9781761170157 (paperback)
        9781761179013 (ebook)
        9781761178115 (ePDF)

Design Josephine Pajor-Markus
Cover design George Saad
Printer Griffin Press

All reasonable efforts were taken to obtain permission to use copyright material reproduced in this book, but in some cases copyright could not be traced. The editors welcome information in this regard.

This book is printed on paper using fibre supplied from plantation or sustainably managed forests.

CONTENTS

FOREWORD: FIRST PEOPLES AND STEM

by Corey Tutt OAM

What does a scientist look like today? I ask this question in classrooms right across the country. But even in 2024, the majority of students still describe a White, middle-aged male wearing glasses and a lab coat.

Our society has taken important steps towards acknowledging and accurately portraying Australia's diversity, including our First peoples. But particularly in STEM (science, technology, engineering and mathematics), there's still much more work to be done. Storytelling, writing and collections like this can help push these fields in the right direction, even if it means simply acknowledging the size and shape of the problem that exists.

When we reflect on STEM, we often find ourselves reflecting on just one side of the STEM sphere. The Western science side. Think Albert Einstein in his lab, surrounded by equations, or perhaps Thomas Edison with incandescent light bulbs flickering to life.

But it's vital to acknowledge that the First peoples of this land have a long history with STEM, well before the term was first used. Traditional knowledge is powerful. It's an important knowledge system, and no lesser than Western science. Aboriginal and Torres Strait Islander people have lived sustainably on Country for millennia. There's evidence of the strength of this knowledge all around us, from structures that were built thousands of years ago to plants being used as resin to create sustainable tools for generations to care for Country. We see that strength carried through in some of this country's greatest minds, like Professor Misty Jenkins, one of Australia's leading brain cancer researchers. First Nations people of this land have always excelled in STEM.

It's important that as a discipline and industry STEM learns from other ways of knowing and being. This includes First Nations knowledge. Slowly, Western science is beginning to understand how valuable traditional knowledge is – and, in some pockets, is starting to see it as a different but equal way of knowing. From traditional navigation to agriculture to sustainable land and sea management, First Nations knowledge systems are adapted over millennia to work in harmony with the natural world, with Country. Crucially, these systems are built around the idea that with great knowledge of the land and its inhabitants comes great responsibility. A responsibility to be stewards of land and community, to be sustainable with that knowledge, to not damage Country or the land, and to respect the owners and custodians of that knowledge.

Writing about Aboriginal and Torres Strait Islander contributions to STEM is not merely an act of recognition – it's a crucial step towards rectifying historical injustices, promoting diversity in knowledge systems and inspiring future generations. At its core, this endeavour serves to weave a more inclusive narrative of scientific achievement and innovation – one that embraces the rich tapestry of Indigenous knowledge and perspectives.

Accurately documenting Aboriginal and Torres Strait Islander contributions to STEM disciplines is one of the most basic steps, and it's an act of reclaiming and preserving cultural heritage. For centuries, Indigenous communities have developed sophisticated understandings of their environments, encompassing astronomy, land management, medicine, and more. These insights, often passed down through oral traditions, represent invaluable contributions to scientific knowledge that have historically been overlooked or marginalised. Properly attributing knowledge to tribes and clans when writing about traditional knowledge, and also allowing the knowledge to be owned by the community and knowledge holders, is vital. Taking this knowledge without permission or acknowledgement undermines First peoples' culture.

Another key step is writing about Aboriginal and Torres Strait Islander STEM pioneers, which serves as a catalyst for inspiring future generations. I am a proud Kamilaroi man who dreamt of working with animals, yet it was really hard to break into the STEM industry due to race and social upbringing. I found the labs to be quite a racist environment. While we have come a long way, the industry still seems to be about 20 to 30 years behind others in terms of its relationship with First peoples. And since the referendum on the Voice to Parliament, the STEM industry has even gone backwards in how it treats Indigenous peoples in the field. By showcasing role models and stories from within First Nations communities, we can begin to undo this damage. We can empower young Indigenous individuals to pursue careers in STEM fields, thereby fostering greater diversity and inclusivity within these disciplines. This representation not only enhances opportunities for Indigenous youth but also contributes to the overall advancement of scientific inquiry through varying perspectives.

Highlighting Indigenous perspectives in STEM sometimes challenges the prevailing Eurocentric narrative that dominates mainstream science. But this is okay. In fact, it's an integral part of science – because science is simply a process through which we come closer and closer to truth. By integrating diverse ways of knowing, we enrich our understanding of complex scientific phenomena and broaden the scope of inquiry.

Importantly, sharing these narratives also plays a pivotal role in reconciliation efforts. Acknowledging the contributions of Aboriginal and Torres Strait Islander peoples to STEM underscores their intellectual prowess and resilience in the face of immense historical wrongs. It signals a commitment to truth-telling and justice, essential components of healing and building equitable futures.

Writing about Aboriginal and Torres Strait Islander STEM achievements also encourages collaboration and partnership between Indigenous communities and scientific institutions at a wider societal level. By fostering respectful engagement and co-creation of knowledge, we can address pressing global challenges with innovative solutions that draw from both traditional wisdom and contemporary scientific methodologies.

But we must also be respectful and careful in how this is done. Many industries and institutions try to own the Indigenous STEM space instead of allowing First Nations people to develop and maintain their own expertise. In the past five years I think we have gone three steps forward and two steps back. Should we be happy with this progress? No, I believe more needs to be done so we can really bring the industry forward to a point where racial inequity doesn't exist – along with gender inequity. These things should be as extinct as the dinosaurs, something we can look back upon and learn from.

Documenting and promoting Aboriginal and Torres Strait Islander contributions to STEM through writing is an ethical imperative and a strategic investment in the future of scientific inquiry. It honours the depth of Indigenous knowledge systems, promotes diversity in STEM disciplines, inspires future generations, advances reconciliation efforts and fosters collaborative partnerships. By amplifying these voices and narratives, we not only enrich our understanding of science and technology but also reaffirm our commitment to justice, equity and respect for all knowledge systems and peoples. Embracing and celebrating Indigenous contributions to STEM is not just a matter of recognition but an essential step towards a more inclusive and equitable society.

It's exciting to see emerging Kaurna/Ngarrindjeri writer India Shackleford wrapping up this book. It's also great to see

Joseph Brookes' piece arguing for why Indigenous science must be a standalone national science priority. But this is just the start. This anthology reflects the state of the broader STEM industry in Australia, which I believe is still a fair way behind other disciplines when it comes to recognising this country's STEM pioneers. We need to decolonise the lab so future generations can see STEM as equitable and inclusive. It's often said, children can only be what they can see, so we need to break the status quo for them. We need to challenge and move beyond unfair bias. We need to celebrate those First Nations people doing this already. We need to hear these voices.

Because we all know it isn't only White middle-aged males wearing lab coats.

INTRODUCTION: VOYAGERS, ALL

Jackson Ryan and Carl Smith

Ours is a country rich with story, ancient and new.

Let's start at the beginning. In a shallow reef, 3.5 billion years ago, microbes began slowly cementing sand and sediment together, forming mineral mats. Today, their fossilised remains are locked in the rock in the Pilbara, Western Australia. Scientists race to read their stories – stories that could reveal life's ignition point.[1]

It was at least 65 000 years ago that the first people arrived in Australia. They took to recording their stories in rock, too – inscribing petroglyphs into boulders and carving symbols in stone with hand tools; running fingers, wet with ochre pigment, along cave walls to create grand landscapes.[2]

Indigenous oral histories connect to the Dreamtime, codifying knowledge, laws, customs, rituals and ceremonies in story.

Oral stories known to be told at least 12 000 years ago pass on vital information to this day. The Palawa people – the Aboriginal people of Tasmania – still speak of the flooding of the ancient land bridge once connecting their island to the mainland. They also describe the position of Moinee, a bright star that once shone near the South Celestial Pole but which has since moved due to the

---

1   'Stromatolites and other early life', WA Department of Energy, Mines, Industry Regulation and Safety, <www.dmp.wa.gov.au/Stromatolites-and-other-evidence-1666.aspx>

2   'First rock art', National Museum of Australia, 6 September 2023, <www.nma.gov.au/defining-moments/resources/first-rock-art>

Earth's tilt. This geological and astronomical evidence shows their stories have remained intact since the last Ice Age.[3]

There's ongoing contention about what humanity's oldest story might be.

Though we certainly shared stories before it, some point to the Epic of Gilgamesh, an ancient Mesopotamian odyssey carved into stone tablets in the Akkadian language, as the world's first piece of written literature.

Twelve tablets tell the tales of the great Sumerian King's pursuit of eternal life. It's considered one of the first stories to detail the 'hero's journey', connecting readers to the personal plight of characters, their transformations in the face of grand challenges. It shows moments of authentic relationships and tugs on your heart, it documents humanity's battle with – and love of – nature, and its narrative arc is as potent and relatable as what we see in contemporary writing.

From those ancient stories to now, one important thing remains the same: the most compelling tales are transportive. They pull us out of ourselves; we land in fully-formed worlds we could not previously imagine. We are explorers, expeditioners, tourists and voyagers, all.

And stories with this quality compel us to share them, too – we pass along our own stories, or those we've heard. Our country's First Nations people did this through song, dance and painting. We've sat around campfires, radios and televisions. Nowadays, we paste a URL in our family chat, or zip it to a friend via Facebook and Instagram. 'Have you seen this?' we say. 'Cracking story!'

In this, we see the echoes of Gilgamesh's most improbable pursuit realised: the closest we can come to eternal life lies not in

---

3   D Hamacher, P Nunn, M Gantevoort, R Taylor, G Lehman, KHA Law, M Miles, 'The archaeology of orality: Dating Tasmanian Aboriginal oral traditions to the Late Pleistocene', *Journal of Archeological Science*, vol. 159 (November 2023), <www.sciencedirect.com/science/article/pii/S0305440323000997>

perpetually deterring death. It lies in sharing stories imbued with knowledge, passing them on to one another, across generations.

It might sound trite but we have been humbled reading the submissions for this year's *Best Australian Science Writing* anthology.

It's been an honour to read the work of our country's best science writers, and it's remarkable to imagine that contained within these pages – which we view as our community's series of stone tablets, arranged neatly for you – is just a single year's worth of storytelling. The best science writing grabs us and pulls us deep into story. It informs. It educates. It surprises. It connects and deepens the human experience – like all great stories throughout human history.

Some start with a bang, like **Liam Mannix**'s descent into the Victorian forest. 'To get to the slime house,' it starts, 'a capable 4WD is highly recommended.' Or **Bianca Nogrady**'s piece, which sets the tone with 'A good espresso coffee is sexy as hell' and then settles, warmly, in your chest as it drinks in the science of coffee. And there's urgency in **Justine Hausheer**'s sprint through the dark sands of the Solomon Islands at night, searching for endangered leatherback turtles to tag with new tech: 'Turtle coming up, turtle coming up. Sector 25' it bellows, signalling the chase is on.

Others are slower burns that build to dazzling brightness. **Amanda Niehaus** achieves the near-impossible, threading the needle of memoir-meets-non-fiction-meets-a-new-dog in 'Dog people', a piece with echoes of Helen Macdonald's masterful *H is for Hawk*. **Kate Evans**' 'Of moths and marsupials' is another form of rare alchemy that pulls threads together into the perfect weave. Sometimes you need depth and breadth in post-mortems like **Elizabeth Finkel**'s 'This little theory went to market', which lets us see our recent history through clearer eyes, with the benefit of distance.

The deep and rich tradition of storytelling feels fresh and fragile in **Dyani Lewis**'s superb piece on rock art. It reminds us that stories can be lost – now matter how long they've already lasted – and so they must be preserved. **Amalyah Hart** and **Ange Lavoipierre** each remind us in their compelling explorations of AI's promises and perils that we are also shaping the course of humanity's story right now and must take care not to lose ourselves along the way. **Drew Rooke**'s fantastic story moves from the computer screen to the future, taking us into the code behind climate models; it reminds us we're writing the story of our planet.

Other stories shine because they fill a page in new and exciting ways. We were absolutely thrilled to include the first excerpt of a libretto in the anthology – an expansive cosmic tale by **Jenny Graves** and **Leigh Hay** that has been performed by the Heidelberg Choral Society.

Plus, from the punchy but personal prose of a pharmacy placement delivered by **Michael Leach** to the playful explosions of **Shey Marque** to the powerhouse that is **Alicia Sometimes**, science poetry delivered again this year, often pouncing off the page.

Many of the finest stories in this anthology also remind us of those we share the planet with. **Zoe Kean** follows the tail feathers of a forgotten emu to a fancy French villa, **Tabitha Carvan** reckons with the possums in her rooftop, coming to something of a truce that inspires introspection and understanding, and **Carly Cassella** listens closely to the songs of liars deep in the forests of the Blue Mountains. Meanwhile, **Natalie Parletta** and **Viki Cramer** both take us from seed to stem in their stories about fauna's steady companions, our trees and crops and plants.

There are those who celebrate the hero's journey. **Cameron Stewart**'s outstanding feature 'Heroes of Zero' shows us how Michelle Haber and the Children's Cancer Institute are slowly winning the fight against childhood cancer. **Angus Dalton**'s rollicking mystery adds another notch to the tally of Australian scientists beating ulcers (alongside Nobel Prize winners Barry

Marshall and Robin Warren, who famously tackled those of the stomach). **Lydia Hales**'s story on body dysmorphic disorder reminds us that many heroes have to fight against the odds to win their war.

Many of the classiest non-fiction stories put the writer in the frame. Facing down a heatwave in WA, **James Purtill** ruminates on the effects of a warming planet on this generation – and the next. **Lauren Chaplin** finds herself in GP clinic after GP clinic on a frustrating longhaul to diagnosis, and **Ellen Phiddian**'s visit to a convenience store sets the tone for her week at a Green Chemistry conference. Bring your brolly.

As always, we are celebrating well-told and reported stories of deep emotion, resilience and triumph too. **Petra Stock** pushes back against anti-math sentiments with carefully cast case studies, **Matthew Ward Agius** surprises by revealing how solar races have sustained Silicon Valley in ways unexpected, and **Rich Haridy** is hypnotising as he follows a psychonaut's trip through the medicine cabinet in search of a solution for long Covid.

Some stories are a mirror for the reader that ask us to reflect on what we see in our lives and societies. **Clare Watson** lays out a compelling case for how we could approach pharmaceuticals based on public good instead of profit. **Joseph Brookes** underlines the importance of prioritising Indigenous knowledge in our country's new science policy framework and **Jacinta Bowler** leaps onto the battlefield as Australian scientists gain ground in the long struggle against governmental meddling in research.

Finally, many stories bristle with moments of delight and entertainment, personality and flair. We see this in **Belinda Smith**'s wonderful exposé on the chemistry of chip flavouring, which leaves the reader satiated and smiling, whereas **India Shackleford**'s buoyant work for childrens' science magazine *Double Helix* shows how the language – and stories – of First Nations people might be taught and preserved in new ways, through the humble emoji.

Before you begin your adventures within these pages, it would be remiss of us – science journalists both – not to mention that science writing is becoming a much more difficult proposition for many. The year 2023 was an extremely difficult one. Major US publishers like *WIRED* and *National Geographic* laid off their science reporters, and the 151-year-old publication *Popular Science* shut its doors.[4] In early 2024, Australian science magazine *Cosmos* made half of its workforce redundant.[5] This mirrors a decline in media jobs more generally, but losing science writers and journalists is a particularly cruel blow – one that can affect the public's trust in science and make it harder to access crucial information to live a healthy, happy, safe life.[6]

That's what makes this collection of stories so powerful. Despite the challenges, our country's science storytellers continue to experiment, grow and find ways to flourish. This is the fourteenth edition of the *Best Australian Science Writing* anthology. Whether you know it, or not, and whether this is your first or your fourteenth anthology, you're a special kind of reader. Your support of these stories is invaluable; you carry the fire that illuminates a pathway for prospective science writers to follow.

As members of the Australian science writing community – a diverse pool of poets, academics, journalists, communicators, reporters, editors and beyond – we can tell you, we are immensely grateful for this support.

If there's a story in here that you love, we hope you'll share it. Post about it. Grab a loved one and proclaim 'You have to read this!'. Track down the author's bio to find more, tweet, email links to your friends, share in your WhatsApp group chats. In this way,

---

4   'Science journalism is shrinking–along with public trust in science', *Science Friday*, 5 January 2024, <www.sciencefriday.com/segments/science-journalism-trust-in-science/>

5   H Karakaluk, 'SA's only science publisher at risk of shutting doors', *InDaily*, 4 April 2024 <www.indaily.com.au/news/adelaide/2024/04/04/sas-only-science-publisher-at-risk-of-shutting-doors>

6   See note 4.

you do what Gilgamesh could not. You give a *story* eternal life. And you encourage the next epic tale to be told.

Go beyond these pages, too. There are many other stories we would have loved to share with you. Others from authors already in the book, many more from writers across Australia who lie just beyond the boundaries of this year's shortlist.

Seek out great science writing like you seek out news and TikToks and YouTube videos. You might find our work near the back pages, or at the end of a news bulletin, on Saturday morning radio slots, or buried within magazines. But every time you tune in or stick around, you remind decision makers that science and stories are both still important.

As you take your first steps into *The Best Australian Science Writing 2024*, passing by the slime house, sneaking along a beach in the Solomon Islands, traipsing through the underbrush of the Blue Mountains and speeding back to the very beginning of the universe, we thank you for carrying the fire, continuing the great tradition of storytelling in our country.

Enjoy your travels.

IN THE HEART OF THE FOREST, ONE WOMAN BUILT A HOUSE OF SLIME

Liam Mannix

To get to the slime house, a capable 4WD is highly recommended. After a 45-minute drive east of Devonport in northern Tasmania, the dirt road peters out into a rocky goat track.

You then face a 2-kilometre trek steadily up Black Sugarloaf mountain, the stringybark and native cherry hemming you in on both sides, before the forest suddenly opens up and the slime house emerges from the gloom.

The walls are mud and local dolerite, and the roof is tin. Tracks lead every which way into the forest. Three or four umbrellas and several saucepan lids are scattered across the clearing.

Inside you'll find a tumble of books, microscopes and grey filing cabinets.

Open a cabinet drawer, and you'll find matchboxes. Thousands of matchboxes. Inside each matchbox lies a wonder. One might hold a collection of fruiting bodies on fragile stems daubed with phosphorescence, reaching for the air. Another is filled with globes of caramel or blueberry. A third contains fine hairlike projections, like sea anemones.

They are slime moulds. This library is certainly Australia's greatest collection outside a research institute, and perhaps the world's.

'It got a bit obsessive,' says Sarah Lloyd, the woman who built this house and collected this slime library.

Under a microscope, a miniature world reveals itself, alien and strange. 'Aren't they beautiful?' says Lloyd.

Slime moulds have gone largely ignored since their discovery. But those who do take an interest often become obsessed. They keep them as pets and feed them oat flakes. They get them to solve mazes, design subway systems, and model computer algorithms after them. They start talking about philosophy and complaining about the human urge to 'put the brain on a pedestal'.

'Sarah just absolutely does not understand why anyone wouldn't be in love with them,' says Dr Margo Adler, founder of Tasmania's Beaker Street Science Festival. Lloyd offered a tour of her property for this year's festival, which sold out almost instantly. 'And then when you spend time with her, you do see how amazing they are.'

In 1869, Oscar Brefeld, an acid-tongued, one-eyed German, was using dishes of agar to coax fungi to spore and flourish from horse dung. When a slime mould emerged, Brefeld wasn't sure what to make of it. They looked like fungus but lacked hyphae, the mouldy growth that consumes food left too long in the fridge.

After several attempts at finding the right spot for them in life's kingdom, taxonomists eventually grouped them with the Protists – leftover creatures that are neither animal nor plant nor bacteria nor fungi. They are Earth's aliens.

Acellular slime moulds begin life as a spore carried by the wind; the spore lands and hatches into a single slime cell. These cells are not special. Like most cells, they can take in food and avoid toxins. But when two cells find each other, they can combine into a single-celled organism known as a plasmodium. With enough food, plasmodiums can keep expanding and expanding in size, their internal components doubling every nine hours, and can move at about 5 centimetres an hour, hunting bacteria.

Apparently, they make very good pets. You can keep them on an agar plate and feed them oat flakes; a treat prompts them to build new tube networks and pulse with happiness. 'I don't know why – they don't eat oat flakes in the wild,' says Associate Professor

Tanya Latty, an entomologist at the same university, who keeps a pet *Physarum polycephalum*, 'but we feed them that and they are very happy'.

Like the slime, Lloyd has gone through many life stages. She's been a writer, a library assistant, made and sold luxury leather goods. When she bought her Black Sugarloaf property in 1988, the bulldozers and chainsaws had just been through, leaving the land a raw wound, the trunks of eucalypts scattered where they fell. There was no house, and she and her partner Ron Nagorcka, a composer, slept in a calico teepee until they could build.

Lloyd has always tracked the animal, plant, insect and fungi life since she moved to Sugarloaf. She records birdsong and photographs hornworts.

Perhaps a decade ago, as she was walking through the rotten logs that surround her home, Lloyd came across several large patches of bright egg-yolk yellow, that had seemingly emerged from nowhere. She later identified them as *Fuligo septica*, also known as scrambled egg slime.

Intrigued, Lloyd turned her camera towards her feet – and discovered Sugarloaf was incredibly slime-rich. There are anemone-like *Ceratiomyxomycetidae*, delicate *Liceales* and flesh-like *Trichiales*. She's found four species new to science and named a new genus.

Lloyd dons a head torch and finds them in darkness. She'll watch them mature for a few days, waiting for the slime to 'fruit', its soft body turning hard, stalks projecting brightly coloured fruiting bodies into the air. If it's raining, Lloyd will erect an umbrella or prop a pot lid over the slime to protect the fragile projections from the drops. She then gathers the bark and leaves and places the sample in a matchbox, which she carefully labels, ready for storage. Or mailing.

The National Herbarium of Victoria, at the Royal Botanic Gardens, is filled floor to ceiling with filing cabinets with floral specimens, perhaps 1.5 million in total. Tom May, the herbarium's mycologist, walks through the towering cabinets, hunting the small alcove that collects everything we know about slime moulds.

May pulls a shoebox from a cabinet at random and opens it. Inside, nearly every sample is marked with Lloyd's name.

'For us, having someone like Sarah, she's out in the bush, she knows exactly what she's doing,' he says. 'It's priceless.'

Most of the work of exploring Australia's plants and animals is done. But slime moulds are uncharted territory. Study them, and you have a chance to see something no human has ever seen before, May says. 'All of a sudden, there's a whole other universe opening up for you.'

A couple of decades ago, Toshiyuki Nakagaki and his colleagues at the Bio-Mimetic Control Research Centre in Nagoya, Japan, sliced up *Physarum polycephalum* and scattered it across a maze. They dropped some food at either end. First, the slime enlarged to fill the entire maze. The researchers then watched, stunned, as the slime shrank down to only the shortest possible path through the maze, avoiding any dead ends.

The brainless blob had solved the maze. 'This remarkable process of cellular computation,' the researchers wrote, 'implies that cellular materials can show a primitive intelligence.'

That was just the start. As it moves, slime lays down a trail of mucus. This allows it to map its local environment and avoid places it has already been. Other research shows slime can quickly learn to anticipate regular changes to its environment, suggesting the presence of an internal clock. And slime can learn to tolerate something it does not like to get food. If you allow slime that has learnt this trick to fuse with naive slime, the slime will share its learning.

Nakagaki would soon turn his slime to real-world mazes: subway networks. He laid out a map of Tokyo's metropolitan sprawl, oat flakes in place of major towns, and then introduced the slime. It quickly built a network of tubes strikingly similar to Tokyo's existing subway map.

Some researchers have turned to algorithms modelled on slime mould behaviours to solve them, trying to capture the way the slime expands everywhere before contracting to a single optimal path. Others are experimenting with neural networks – the technology that underpins AI – built in ways similar to how slime strengthens useful tubes and prunes away those that are no longer needed.

In her lab, Tanya Latty has given slime moulds a selection of foods placed under different levels of light and darkness. The slime, which hates light, happily ate all the food in darkness – but only its favourite foods under lights. It was making trade-offs. 'It does not like taking a risk – unless it is worth it,' says Latty. 'That one surprised me.'

No one really knows how slime does all this. The best theory researchers have is slime waves. Each region of slime has a pulse, like a localised heartbeat. When a slime region senses something good – like food – it speeds up its oscillations. These oscillations flow through the entire slime as waves, allowing distant parts of the organism to communicate. Tubes are remodelled in the direction of the food. The rest of the slime shifts over to investigate.

Lloyd says this line of scientific inquiry will continue to illuminate the wonders of slime. 'We're only really just appreciating now how important fungi are,' she says. 'And slime moulds, I think their day will come.'

An individual human neuron is no more capable of solving a maze than an individual slime cell. But by coordinating their rhythmic electrical activity – brain waves – individual neurons combine into intelligence more capable and complex than any computer built so far.

The development of human civilisation is enabled by a wilful ignorance of the intelligence of animals – it's what allows us to exploit them so horribly. We treat plants and fungi and protists with even less respect, swathed in the comforting belief these creatures can't think or feel.

Slime moulds pick at that logical inconsistency, by asking: what makes us so special?

✳ *Dog people: How our pets remind us who we really are*, p. **40**
How scientists solved the 80-year-old mystery of a flesh-eating ulcer, p. **145**

SATELLITE TRACKING THE PACIFIC'S MOST ENDANGERED LEATHERBACK TURTLES

Justine E Hausheer

'Turtle coming up, turtle coming up. Sector 25.'

I'm sitting on a driftwood log, watching waves crash against the moonlit shore when the walkie-talkie crackles. It's just after 9.30 pm; we've only just started our shift patrolling the beach for nesting leatherback sea turtles, and the call comes sooner than we expected.

Rick Hamilton, director of The Nature Conservancy's Melanesia program, grabs the walkie-talkie. We wait, holding our breath with anticipation. 'New turtle!' shouts an excited voice through the radio. 'Hurry, she's already digging.'

We hurtle back to the thatch shelter to grab our bags. Mine full of camera gear, Hamilton's holding a work light and three satellite tags cradled in an old shirt. And then we run.

Our feet sink into the soft black sand, waves slapping against our shins as we search for stable footing on the sloping beach. Lights will spook the turtle, so we run in darkness, navigating around massive driftwood logs and stumbling over vines and coconuts.

Just when I think I can't run any farther, I see a thick, dark line perpendicular to the waves. Another log? No. It's where the turtle crawled ashore, leaving behind a wide, deep track, like the tread of a monster truck. We're here.

The other rangers are waiting in the forest's shadows, all gazing intently at the massive lump against the tree line. We join them, gulping for air and dripping sweat. From the darkness I hear flippers thwack against palm fronds, the swoosh of sand spraying, a soft snort.

A ranger creeps forward, quickly shining a red-light torch to check her progress. The narrow beam illuminates her backside, flippers scooping sand out of a deep, narrow chamber. 'She's ready!' a voice whispers.

We flip on our headlamps, throw down our bags, and then chaos ensues.

Nature Conservancy scientists are working hand in hand with conservation rangers from Haevo community, in the Solomon Islands, to attach 10 satellite tags to nesting leatherback sea turtles. The data they gather will help protect some of the most important nesting beaches for these critically endangered turtles, and reveal their migration routes across entire oceans.

Last chance for leatherbacks

My first encounter with a leatherback came just a few days earlier, at the beginning of our week-long stint at Haevo. She crawled up from the sea just before moonrise, digging her nest amidst a tangle of vines on the high dune.

A tape measure stretched across her carapace reads nearly 1.8 metres long, and her front flippers extended another 76 centimetres on either side. Hard and rigid as airplane wings, those flippers enable her to swim at speeds faster than a bottlenose dolphin, or to dive to depths of nearly 1220 metres on a single breath. She spends most of her life in the open ocean, feeding on jellyfish far from shore.

Leatherbacks' evolutionary origins stretch back more than 100 million years, but in the past few decades their populations have declined dramatically. The turtles in the Western Pacific are amongst the hardest hit. Scientists estimate that this genetically distinct subpopulation is now critically endangered, having declined 83 per cent in just three generations. Only 1400 breeding adults are thought to survive, and the population is still falling.

By 2040, scientists predict that the Western Pacific leatherbacks will be whittled down to just 100 nesting pairs each year. 'They're crashing hard, and it's going to continue unless we arrest the decline,' says Peter Waldie, acting director of TNC's Solomon Islands program.

Western Pacific leatherbacks nest across Indonesia, especially West Papua's Bird's Head Peninsula, Papua New Guinea, and the Solomon Islands. In the Solomons, they concentrate at just a handful of key nesting beaches, including the roughly 3-kilometre stretch of black volcanic sand known as Haevo.

It's just one beach, but what happens here could tip the balance for the entire population of Western Pacific leatherbacks.

A night on patrol

Earlier that evening, the rangers start work at sunset. A dozen men and women dash about the thatch-roofed hut, changing into their bright green work uniforms, brushing teeth, and gulping down the last of their rice-and-tuna dinner as some indistinguishable local pop music plays on the radio. The plastic folding tables are scattered with charging flashlights and two-way radios, spare pens and data sheets, and thermoses of instant coffee and tea ... everything we need for a night's work on the beach.

As we get ready, I ask the rangers what Haevo was like before the conservation work started.

'During the nesting season, the men from the village would come to the beach and wait,' says Benson Clifford, who has worked as a ranger here for eight years. 'When the leatherbacks came ashore, they would kill them and share the meat with the entire village.'

Sea turtles are a culturally important animal to Solomon Islanders, woven through their history and kastom. But they're also protein. Prior to starting work as rangers, all of the men and women I spoke with ate turtle meat, typically on special occasions like weddings, birthdays or funerals. When Hamilton and I journey to Haevo on an overnight local ferry, the boat captain announced that the ship rules included: no smoking, no drinking, no drugs, and no live turtles. Even in Australia, shoppers could find canned turtle meat on the supermarket shelves right up into the 1960s.

In the past few decades, human population growth has increased the hunting pressure to unsustainable levels. Combined with other threats, like entanglement in longline fisheries and rising seas washing away nesting beaches, the pressure is pushing leatherbacks over the brink.

'You don't have to be a rocket scientist to realise that if you kill all the turtles on a nesting beach, you're in trouble,' says Hamilton. 'These animals migrate across entire oceans, but when they come here to nest, it's a critical, life-stage bottleneck. That gives us a real opportunity to protect them when it matters most.'

Leatherbacks are the only turtle species completely protected under Solomon Islands law, but the rangers tell me that most people aren't aware of the rules and enforcement is nonexistent. Situations like this are common in countries where resources for educating fishers and enforcing regulations simply aren't available. Few, if any, of the rangers here knew it was illegal to kill a leatherback until the monitoring program started.

Other turtle species can be hunted for subsistence, not sale, but they can't be killed while on the nesting beaches. In reality, both rules are almost never enforced. A previous TNC study surveyed local markets for turtle products, finding thousands of turtles for sale.

'All of the turtles that came up to nest would die,' says Clifford. 'And they kept killing nesting leatherbacks right up until 2013.' That was the year that TNC started a leatherback monitoring program at Haevo.

It's fully dark by the time we leave. Slipping off our shoes, we walk along the edge of the waves. Our destination is the ranger station at the far end of the beach, which curves gently around a small bay to the mouth of a crocodile-filled river. The black sand still radiates the sun's warmth, and banks of steam envelop us as we trudge.

We arrive 30 minutes later. The rangers disperse in groups of two and three, each taking up their vigil on a different section of beach. Hamilton and I stay just outside of the camp, watching shooting stars streak across the sky and scanning our section of beach for any movement.

The rangers make this trek each and every night, watching over the beach and recording data on any turtles that come up to nest. Similar monitoring programs are underway at Sosoilo, an hour's boat ride north of Haevo, and at Sasakolo, on the island's western side.

Hamilton says that the leatherback work is modelled off of similar work in the Arnavons, where TNC has worked with local communities for nearly three decades to protect a critical rookery for hawksbill sea turtles. Nesting numbers there have more than doubled in the past 20 years, and the Arnavons was recently named the first marine national park in the country.

The leatherback pit crew

Hamilton and I are still catching our breath as the rangers swarm around the nesting turtle like a pit crew around a race car. She's entered a trance-like state as she starts to lay and barely registers the frenzy of activity. But we only have about 20 minutes.

Her eggs drop by twos and threes into the deep cavity in the sand. Rangers Nora Tuti and Lonsdale Balu gently remove them, counting as they go, and place the eggs into a 19-litre plastic bucket. They'll be re-buried safely in a hatchery in just a few minutes.

Rodney Heinz crouches down to attach two metal tags, one to each of the turtle's hind flippers. Then he moves around to face the turtle and takes a small DNA sample from one shoulder and inserts a small microchip into the other. (Similar to the microchips used to help identify lost pets, it provides a backup identification method to the flipper tags.)

As soon as Heniz finishes, Lynette Haehathe pulls out a yellow tape measure and stretches it across the turtle's shell from tail to shoulder. At 160 centimetres, she's only a medium-sized leatherback. Even so, she likely weighs more than 450 kilograms. Jessica Haraputti stands ready with a small notebook, recording the tag numbers and carapace length.

At the centre of the chaos, Clifford and TNC's Simon Vuto work quickly to attach the satellite tag. Held on with wire thread and epoxy, the tag will record the turtle's location continuously, and then transmit that information to nearby satellites. The tags can also record sea temperature, how deep a turtle dives, and when turtles leave the water to nest.

Thanks to the monitoring program, Haevo's leatherbacks are well protected, and nesting numbers are starting to increase. But there's no guarantee that this turtle will return to Haevo in ten days' time to lay her next clutch of eggs. Unlike other turtle species, leatherbacks aren't strictly site-loyal. They might nest on their natal beach, where they themselves hatched, or they might nest at another site nearby.

With so much time and effort going into protecting Haevo's leatherbacks, Hamilton and his team want to make sure that these turtles are well and truly protected, even if they choose to nest elsewhere on Isabel Island.

The satellite tag we're attaching tonight is one of ten tags being employed during this nesting peak, which stretches from November to January. The rangers will attach more tags in a few months' time, during the smaller mid-year nesting season in May and June.

The data from those tags – which records the turtle's exact GPS locations – will reveal if any of Haevo's leatherbacks are nesting elsewhere on the island, and can help guide the establishment of future community conservation areas.

'If a turtle nests at an unprotected beach, there's a good chance that her eggs could be inundated with saltwater or she will be killed for food,' says Hamilton. 'So our goal would be to create a network of ecological linked protected areas.'

Once the turtles are done nesting, the tags will track them along their months-long journeys to their cold-water foraging grounds.

'Earlier work suggests that most of these turtles – the ones nesting in the austral summer – will go back to foraging grounds in New Zealand and Tasmania,' says Hamilton. 'But we suspect that some of the turtles that nest here, in the middle of the year, migrate more than 10 000 miles from the US west coast.'

A journey of that length represents one of the longest migrations in the world, on par with larger pelagic species, like humpback and grey whales, or bar-tailed godwits travelling from Alaska to New Zealand.

The journey home

The rangers finish just in time, stepping back as she starts to fill in the now-empty nest cavity. She wiggles back and forth, front flippers sweep sand backward as her hind flippers push the piles into the hole. The satellite tag looks like a tiny barnacle against the smooth curve of her carapace.

Benson is already walking down the beach towards the closest hatchery, where he'll re-bury the eggs above the high tide mark.

I kneel down in the sand in front of the turtle, gazing into her eyes one last time, my face inches away from her own. She blinks, and the clear mucus dripping from her eyes jiggles as she stares back. And then, by way of goodbye, she snorts in my face.

I wake with a start as Anita Posala gently shakes my foot. 'It's time to go,' she calls, few traces of exhaustion on her cheerful face.

It was a one-turtle night, and I'd managed to get a few hours of sleep. We gather our things and start the long walk back to the ranger's station, looking back over our shoulders to watch the sun slide up into the salmon-pink sky. The rangers call this the 'zig-zag walk'; we all stumble down the beach, so tired we can't walk in a straight line.

Forty minutes later we reach the station. I'm brewing a cup of tea as Hamilton calls me over, motioning to his laptop. On the screen, a map of the Solomon Islands, all deep green islands scattered across dark blue seas. And offshore of Haevo, a bright pink line, heading out to sea.

Weeks later we'd watch that same track, along with nine others, as the leatherbacks finished nesting and began migrating back to their foraging grounds. Some swim south-east, passing Fiji, others turning more sharply south to beeline for New Zealand, or perhaps Tasmania. Three swim straight through a cyclone.

Each alone in the open sea, carrying the future of their species on their backs.

✱ *Dog people: How our pets remind us who we really are*, p. **40**
The last King Island emu died a stranger in a foreign land,
   p. **193**

WESTERN AUSTRALIA HAD ITS HOTTEST SUMMER EVER, BUT CLIMATE CHANGE BARELY MADE THE NEWS

James Purtill

In early February of this year, when the Perth heat really got going, my partner was 28 weeks pregnant.

On the day of the midwife appointment, we snuck to the car from the one air-conditioned room where we both worked, huddled in the cool. Windows up and AC on full, we drove to the hospital.

The midwife told us all the usual pregnancy things to keep the baby safe, but it felt odd to be learning about diet and sleeping positions while outside the city baked and trees died.

We read scary studies showing the impact of extreme heat on pregnant women and unborn babies. We read that extreme heat kills more people in Australia than all the other natural disasters combined.

This has been Western Australia's hottest summer on record – and the hot weather isn't over yet.

Speaking as someone who grew up in WA, the recent heatwave was brutal and unprecedented. The cooling sea breeze stayed out to sea and some nights it was over 30 degrees Celsius at midnight.

For 24 hours in mid-February, the 15 hottest places in the world were in WA.

At an after-work drinks event I attended, a man fainted and collapsed. We gave him water and walked him to his car, through the city. The baking hot streets were utterly deserted.

As the month progressed, there appeared to be a growing disconnection with the way news outlets were generally covering the ongoing natural disaster. News stories often showed people 'beating the heat' by going to the beach. A prominent politician devoted one sentence of their weekly column to the weather: 'Yes, it's summer, and yes, it's hot.'

Richard Yin, a Perth GP and deputy chair of Doctors for the Environment, said the lack of acknowledgment in the media about the impact of heat and climate change was 'vaguely terrifying'.

'Everything is being normalised, as though it's just another heatwave … What we see now is a harbinger of what's to come. This is not even the new norm, it's the lowest level of the new norm. What we're expecting is much, much worse.'

The perception that the media is downplaying climate change is more than a gut feeling. It's backed up by data.

Stories more likely to mention cricket than climate change

Detailed analysis of media coverage performed exclusively for the ABC by Monash University researchers, showed fewer than one in 20 stories about the WA heatwaves mentioned climate change. About one in five referred to the health impacts of extreme heat.

The vast majority of the 172 stories about the WA heatwaves mentioned neither climate change nor heat-related health problems. There were more than twice as many stories about how the heat could affect the result of a cricket match in Perth in mid-February than about how climate change was driving the heatwaves.

By comparison, about half of the stories about the 2019–20 Black Summer bushfires mentioned climate change, Tahnee Burgess, a researcher at Monash University's Climate Change Communication Research Hub, said.

'And if we are seeing discussion of climate change around the WA heatwaves, it's very infrequent and often in passing,' she said.

'So you're not necessarily understanding the dangers and challenges of this warming world and the heatwaves that it can drive.'

This trend wasn't only seen in WA, she added. The analysis also found about 9 per cent of the 528 articles about heatwaves across Australia in February referred to climate change.

'We can say that, overall, the reporting of climate change and heatwaves for the month of February in Australia was significantly lower than we've seen for other extreme weather events,' Ms Burgess said. 'These figures are really, really low, especially considering we also had a heatwave across massive parts of the country. February really was a month of extreme weather, so you'd think it would be a good month to be talking about climate change.'

Report linking heatwaves to climate change totally ignored

Heatwaves are one of the most direct and well-observed consequences of a changing climate. Climate scientists in Australia and overseas have been repeating this point for more than a decade. But despite the solid science, news outlets appear generally reluctant to communicate this to readers.

In early February, as WA was entering a four-day heatwave, the Climate Council published a report and a media release underlining the link between climate change and the summer's extreme weather, such as floods in the east, fires in the north, and heatwaves in the west. Although the report made clear the hot weather in WA was a result of climate change, not a single news outlet published this information that month, the Monash analysis showed.

A second report, published the following week, warned about the health impacts of WA's extreme heat while reiterating the link between the heatwaves and climate change. Of the seven stories that referred to this report, most completely stripped out the reference to climate change, Ms Burgess said.

So what is the link with climate change?

Simon Bradshaw, director of research at the Climate Council, said the Australian landmass as a whole has warmed by about 1.5 degrees Celsius since 1910, making heatwaves worse and more frequent.

'Everything we see today is on a planet made hotter by the burning of coal, oil and gas.'

Perth is warming faster than most other areas. The average summer temperature recorded at Perth airport has increased by about 3 degrees Celsius since 1910, well above the national average.

'We see a particularly strong warming trend in some parts of WA, including around Perth,' Dr Bradshaw said. 'That means longer, hotter, more intense heatwaves. And we clearly had a strong taste of this over the last summer.'

Climate change is making parts of the city 'unliveable'

Growing up near the beach in Perth, I loved summer. Yes, it was hot, but I quickly learned to not complain about the heat. Whingeing marked you as an outsider. The best strategy seemed to be stoic acceptance of plus-40-degree days, combined with giddy celebration of the beach and everything it offered.

Summer shades into autumn, the rain comes, and you soon forget the worst of it. Six months later, the heat starts up again.

Heat is normal, but the orderly progression of hotter summers is now making parts of the city dangerous for some residents.

'The city is becoming unliveable for those who are vulnerable,' Dr Yin said. 'I can understand the need to not scare people, but at the same time we need to be informing them.'

By WA standards, Perth's outlook is relatively cool. A peer-reviewed study published last year found Broome and Port Hedland risked being generally uninhabitable within 70 years. The northern Australian towns are on track to have more days per year over 35 degrees Celsius than under by 2090, Climate Council modelling shows. For Perth, the figure is closer to about 40 days per year by 2090 (up from an average of about 22 before 2010).

But even this relatively temperate climate may be too much for those who are most vulnerable to heat, including the elderly, children, babies, pregnant women, and people with chronic illnesses.

A project run by Better Renting found temperatures in many Perth rental properties exceeded the World Health Organization's guidelines during one of the heatwaves of the recent summer.

The Western Australian Council of Social Service (WACOSS) has begun mapping temperature variation across the city and plans to build public cool-space sanctuaries for residents.

'The dangers posed by extreme heat still aren't being adequately addressed,' WACOSS CEO Rachel Siewert said. 'If you don't have access to air-conditioning and if your building [isn't] energy efficient, you're at risk of high heat. We've heard of people taking shelter in shopping centres and libraries.'

'Insidious' heatwaves kill silently

Why do we routinely underestimate the danger of heatwaves?

One answer to this is the 'insidious' nature of the events themselves, which makes it harder to precisely attribute mortality, Sarah Perkins-Fitzpatrick, a climate scientist at the Australian National University, said. 'We know heat kills but it's hard to quantify by how much. People might die of cardiac arrest or renal failure, so it's very hard to track the precise [mortality rate].'

Deaths are seldom attributed to the heatwave itself, and the immediate cause of death, such as diabetes or a heart attack, is usually reported as the reason for hospital admission. It can take months of combing through coronial records to determine how many people may have died from a specific heatwave.

Heatwaves are also invisible. Unlike a bushfire or a flooding river, extreme heat doesn't make good TV.

And perhaps, along with these reasons, there may be another one.

Dr Yin, who's also president of the Conservation Council of WA, said WA was lagging behind other states in responding to climate change, despite being particularly exposed to its impacts.

'It's completely jarring,' he said. 'We're one of the most vulnerable places in the country for climate impacts.'

Even as the north gets too hot, the supply of arable land shrinks, the forests of the south-west dry out at a record rate, and birds drop dead from the sky, WA has been slow to act, Dr Yin said. It's the only state without an emissions reduction target for 2030 and the only state to have increased its emissions since 2005.

It's also one of the world's largest exporters of natural gas, a fossil fuel that contributes to climate change.

The WA government has, however, committed to achieving net zero by 2050 and in February 2024 released a plan outlining the steps it will take to reduce emissions and help communities adapt to climate impacts including extreme heat.

The climate adaptation strategy includes mapping heatwave risk areas across WA and building greater public awareness of climate risks and climate science.

Preparing for the next summer's heatwaves ... and the next

On the very hot days in February, as the temperature climbed, we cranked the air-con and hoped there'd be enough power. Our unborn baby grew to the size of a sweet potato, according to a website. The next week it was a mango. The next, a coconut.

Towards the end of the month, yet another study came out showing extreme heat in the third trimester was strongly associated with increased preterm birth risks. Heat-stressed mothers gave birth early.

Every morning we shut up the house and checked the weather for a cool change. We ran the air-con so hard a weed grew in the pool of its condensate water.

When the cool change finally arrived, the feeling was blissful. But the sea breeze also brought the scent of smoke from bushfires to the south. Maybe a week later, we had rain.

And for the first time, I found myself dreading the next summer.

✱ *Why solar challenges? They're in the DNA of Tesla,*
    Google for starters, p. **91**
    Predicting the future, p. **122**

THE HEROES OF ZERO

Cameron Stewart

Vivian Rosati fingers the cross around her neck and speaks of faith and miracles. She is at the kitchen table of her home in Sydney's North Ryde when her 15-year-old son Jack bounces home from school with a grin.

He's hatching a plan to go mountain biking with his mates. 'It's his fave at the moment,' she says. Jack then reels off a list of his other faves, from punching his boxing bag to cross-country running, skateboarding and surfing. 'Any sport, really,' says Jack as he leaves the room to change out of his school uniform.

When he is gone Vivian falls silent and tears glisten in her eyes. 'I'm so glad that mine is a happy story,' she says finally. 'There are going to be a lot more happy stories like mine in the years to come. But Jack is my miracle.'

Five years ago, Vivian, a single mum, was watching her only child dying of cancer in front of her eyes. A year earlier Jack had been diagnosed with a benign brain tumour; then one morning he woke up vomiting and with a terrible headache. The tumour had returned – and this time it was cancerous and had spread to his spine. Jack was soon bedbound, unable to walk, eat, or see out of his right eye. It looked like he wasn't going to make it.

'A group of medical staff took us into a room and gave us devastating news about Jack's condition,' she recalls. 'I remember sitting and looking down at the floor, not being able to face these doctors and professors – but my eyes were flooded with tears, I couldn't see anything anyway.' The doctors also told her that as part of a new type of possible treatment, Jack's biopsy was going to be examined by a fledgling program known as Zero at the Children's Cancer Institute in Sydney, headed by a professor called Michelle Haber. 'I just thought, "Yeah great, this program might save kids in the future but my kid is dying now",' says Vivian.

But then three astonishing events unfolded together. Firstly, after three weeks of tests, the Zero program identified the specific genetic mutation – called BRAF V600E – that was driving the cancer growth in Jack. Secondly, there was a drug already available to target that rare mutation. And thirdly, from the moment Jack took the drug, he literally rose from what would have been his deathbed. 'He regained his appetite, colour returned to his face, his headaches and body pain subsided,' Vivian recalls, as more tears are triggered by the memory. 'A couple of weeks on he decided to pick up a tennis racquet and was playing tennis for the very first time. Not long after that, he was selected to represent his school team for cross-country.

'We were stunned – everyone was, from the neurosurgeon to the oncologists,' she says. 'They could not believe it was the same kid.'

Vivian says that Jack's miracle recovery was like having the winning lotto ticket. I ask her who she thinks helped to give Jack that lotto ticket. 'Michelle Haber,' she replies. 'An incredible woman: smart, determined and unstoppable. She is the power behind this fight.'

Professor Michelle Haber is hardly a household name to most Australians, but speak to any parent who has experienced the trauma of having a child with cancer and they will almost certainly know who she is.

Haber, who for the past 20 years has been executive director of the Children's Cancer Institute – the only institute wholly dedicated to the search for a cure – has become the public face of the fight against children's cancer in Australia. But what makes her story so remarkable is that she and her institute are slowly but surely winning this historic battle. They are leading the world in new treatments and giving a new generation of parents and kids hope when once there was none. This is not a fast-sell medical spin offering false hope; the progressive breakthrough in the fight against childhood cancer is a reality backed by hard data.

'Back in the 1950s and 1960s, cancer was virtually a death sentence for children,' says Professor Murray Norris, who has worked with Haber for more than 40 years. 'It really didn't matter what kind of cancer they had. But today the overall survival rate is up around 84 per cent, so we've come a long way over time. And we are absolutely at the forefront of what is being done internationally.'

Yet the story of the fight against children's cancer in Australia has been far harder than these encouraging statistics might suggest. It has been a hardscrabble and often heartbreaking struggle, driven initially by desperate and good-hearted parents and backed by a small team of dedicated scientists. Progress has often been cursed by funding shortages and by lopsided victories where one type of children's cancer is all but conquered while other types are still certain killers.

And despite these huge strides, cancer is still the leading cause of childhood death from a disease in Australia, with around 1000 children and adolescents diagnosed each year, of which just under 200 will die. But Haber is the general who, with her team, is changing the course of this battle. She is masterminding what she believes will one day be a complete victory over children's cancer. In doing so, she has caught the attention of the medical world.

'She is incredibly well known globally,' says Professor Andy Pearson, a former professor of paediatric oncology at Cancer Research UK, the world's largest independent cancer research organisation. 'Michelle Haber is one of the leading scientists in children's cancer research in the world. She has led many innovations in research and one of her great strengths is that her research has a very direct and immediate benefit for children with cancer,' he says.

Haber certainly doesn't look like the stereotypical nerdy scientist. On the day we meet in her office at the Children's Cancer Institute she is dressed immaculately, with a pearl necklace and earrings. She speaks quickly and confidently, like someone who knows her subject backwards. Within minutes you get an inkling why so many parents of sick kids describe her as 'a force of nature'.

'When I look at how far we have come, it makes me so proud and I think to myself, "What a journey",' she says. 'And yet every time I hear the stories of the parents and the children who are still not making it, I realise that we still have such a long way to go.'

Jack Kasses remembers the days when there was nowhere to go for parents when their children were diagnosed with cancer. 'My little girl Helen was put into that building down there,' says Kasses as he stands on his balcony in Sydney's Little Bay and points at an old building which was once a part of the former Prince Henry Hospital. He has never moved from Little Bay since he rushed his sick daughter there in 1975. 'She was just like a lifeless body lying in bed and I was convinced she was going to die,' he says. 'She was just six years old.'

In the 1970s the Kasses family were running several shops near Bundaberg in Queensland when Helen suddenly became ill. 'I didn't even know that kids could get cancer and I had no idea that there was a thing called leukaemia,' says Kasses, now 83. They took Helen to Sydney to seek the best possible care in the hope that somehow she might pull through.

'In those days nearly all the children with cancer died, and in the same week that Helen was diagnosed, a little boy called Robbie was also diagnosed,' Jack recalls. He and Robbie's father, John Lough, would hang outside the hospital together, talking. They became close friends. One day they had a conversation that would change the outlook for all children with cancer in the country.

Jack recalls: 'John said to me, "We've really got to help these doctors. We've got to help them to do more research, to get a more scientific understanding of what children's cancer is and how to cure it."'

The two fathers were not rich but they resolved to launch a fundraising drive to get money to set up Australia's first children's cancer research laboratory. Both were members of Apex clubs, and they pushed hard to harness the resources of the Apex national network to raise funds around the country for their vision. The campaign attracted the support of advertising guru John Singleton, who devised the campaign for free with the slogan: 'Some kids make it, some kids don't. Help a kid make it.'

The slogan was all too true. Jack's daughter Helen did make it, and today works as a teacher in Sydney, although she still suffers side effects from treatments she received. But John Lough's son, two-year-old Robbie, succumbed to his cancer. The two dads forged ahead regardless. The fundraising campaign for kid's cancer became a juggernaut, raising an astonishing $1.3 million – close to $8 million in today's money.

In 1976 the money raised by the two fathers led to the opening of the Children's Leukaemia and Cancer Foundation in Sydney, the body that would later become the Children's Cancer Institute. 'What separated this laboratory from almost any other globally was that it was built not by governments seeking re-election or by philanthropists seeking recognition but by families impacted by the disease,' says Haber. 'And we never ever felt that the unit was anything other than a dream and goal of those two dads and the others that joined it.'

Eight years later, in 1984, the centre appointed its inaugural post-doctoral scientist. 'We had no idea on that day that we had also appointed a visionary – someone who could help make our dream of helping defeat childhood cancer a reality,' says Kasses. 'But we had. Michelle Haber turned out to be that person.'

There was no pre-ordained pathway to the career that Haber eventually chose. There were no doctors in her family; her Jewish parents were ten-pound Poms who emigrated from Liverpool when she was five years old. Her mother raised three children while her father, a linguist, became vice-principal of Mount Scopus College in Melbourne and then the principal of Moriah College in Sydney.

'It was never overtly stated, but our parents had great expectations of us,' recalls Haber. '[They thought] if we worked hard and applied ourselves we would be fabulous.'

Haber studied English, French and Hebrew in her final year of school, as well as maths and science. She deliberately chose school subjects she thought would be fun rather than useful because she imagined that, like her mother, she would soon get married and would have to give up her career to raise a family.

But that script didn't pan out. She chose to do a clinical psychology degree at the University of New South Wales, winning a University Medal for it, before swapping to start a PhD in animal learning behaviour. Yet none of this captured Haber's imagination. She was at a loss about what to do. 'I thought, "Can I do something that will make a difference?"' she recalls. One day she was staring at the list of floors in a hospital when she saw the word pathology. 'I thought that sounded interesting so I literally went up to the fourth floor and I said, "Is there someone I could speak to about doing a PhD here?"'

At her first lecture of the pathology course, she sat down next to a third-year student named Paul Haber. That was in 1979. They have now been married for 42 years, and have three children. Paul currently heads the Addiction Unit at Sydney's Royal Prince Alfred Hospital. Michelle Haber by then had decided that medical research was going to be her field, but what sort of research?

She and Paul went to Israel for three months in 1982 where she did research into molecular virology in Jerusalem. While she was in

Israel she was contacted by her thesis supervisor, Professor Bernard Stewart, who said he was going to be the inaugural director of the fledgling Children's Cancer Institute in Sydney and would Haber be his first scientist?

Just after she returned to Sydney, Haber experienced a day that changed her life forever. 'There was an official launch of the new laboratories and oncologist Professor Darcy O'Gorman Hughes was giving a speech. He said, "Would all the children who have survived childhood cancer please stand up."' As Haber recounts the story, her voice wavers and tears well in her eyes. 'Then this forest of children stood up. And I knew that 15 years earlier none of them would have been alive. But they were alive, and they stood up, and I thought, "This is the future. This is going to be my career."'

Haber started at the fledgling institute in 1984 alongside two other scientists, Professor Murray Norris and Professor Maria Kavallaris. All three of them have now worked together for 40 years, pushing and probing for breakthroughs.

The first came in 1996 when the institute's researchers made the world-first discovery that the treatment of a type of children's cancer called neuroblastoma was failing because it was linked to a drug-resistant protein. This opened the door for a host of new treatment strategies.

Then in 1999 the institute developed a huge breakthrough in the treatment of the most common childhood cancer, acute lymphoblastic leukaemia (ALL). One quarter of all children with ALL were relapsing after treatment – but because it was impossible to know which child would relapse, all kids received high levels of chemotherapy and radiotherapy, leaving many of them with lifelong side-effects such as heart disease and infertility.

Haber's team asked themselves, what if there was a way to predict which kids would relapse and which would not? They worked with oncologists at the Kids Cancer Centre at Sydney

Children's Hospital to develop a test that could identify the risk of relapse, allowing those kids who were at high risk of relapse to receive intensive therapy earlier while sparing those kids who were not at risk of relapse. In a ten-year trial that ran until 2011, the survival rate for kids with high-risk ALL doubled from 35 per cent to 70 per cent. 'I remember thinking when I saw that figure that we were really making a difference and that the motto – "To cure every child with cancer" – was no longer aspirational but possible,' Haber says.

Then in 2005, one of the institute's scientists, Professor Richard Lock, developed the first patient-derived animal models for cancer, known as the PDX model, in which physical samples of a child's actual cancer are grown in a mouse model, allowing scientists to quickly test different drugs and treatments on them to see which might work. Haber says this was a 'game-changer' because it allowed for unlimited testing of new therapies for children's cancer.

In the most crucial breakthrough so far, Haber read a scientific paper in 2010 that described a small study of 70 adults with cancer where the scientists tested their genes for responsiveness to particular drugs. 'Knowing that the challenge in finding the right treatment for every child is that every child and every cancer is different, I thought we could do this with kids and we can do this better,' says Haber. 'I got the team together and said, "What if we can test every child with cancer for cells that express a high sensitivity to a particular drug – effectively finding a drug that will specifically kill that child's cancer and only that child's cancer?"'

Haber's idea was a eureka moment for the field of childhood cancer, allowing for the genetics of tumours to be tested for clues about what is driving the cancer in that particular child and how it might best be treated. In other words, every Australian child with high-risk cancer was given the opportunity to enrol in a personalised medicine program to study the genes of their particular cancer, with the aim of identifying a targeted treatment. Using this method the institute in 2015 launched the Zero Childhood Cancer Program –

the one that saved the life of Jack, Vivian Rosati's teenage son. The program now includes every children's hospital in Australia as well as 22 national and international research partners.

In 2017, a three-year national clinical trial for the Zero program opened; it found that it was possible to explain the molecular basis of 90 per cent of cases of children with high-risk cancer, opening the way for more effective drugs to be developed. The findings, published in the scientific journal *Nature*, made news in cancer research centres around the world.

'Of the first 250 children with high-risk cancer who were analysed on the Zero program and then received the recommended personalised treatment for their genetic change, 70 per cent of them either had a complete or a partial remission, or the growth of their tumour had stabilised,' says Haber.

But these hard-won gains in the battle against childhood cancer are uneven. Some kids just never get a fighting chance.

Mary-Ellen Rogers ushers me into her Sydney home past a series of painted rocks in the garden. One written in purple paint says: 'Amity we love you.' Another reads: 'Funny and caring.' In the garden is a brass plaque that reads: 'Amity Margaret Rogers: Our funny, joyous, clever, loving, spirited daughter ... desperately missed by her "best family forever", wrapped in eternal joy, our darling Amity.'

Amity was almost five years old when her parents, Mary-Ellen and Jackson, noticed that she was no longer her usual bubbly self. Her behaviour became more erratic but they thought the problem was psychological rather than physical until one day a doctor suggested that Amity have a CT brain scan. 'The doctor came back to me and said, "We need to talk,"' recalls Mary-Ellen. 'Then he wanted it to be in a private room, then he said, "I need your husband to bring a support person." So I knew. I thought, "Right, I get it, she's going to go."'

Amity had the most aggressive childhood brain cancer possible. Known as diffuse intrinsic pontine glioma, or DIPG, the tumour occurs in the brain stem and is always fatal. It kills around 20 Australian children a year.

For the next 18 months, as Amity's body slowly stopped functioning, her parents and her two brothers did whatever they could to enjoy the time they had left together. 'Who knows what is wrong and what is right to do in that situation, but we all just tried to live as normal a life as possible,' Mary-Ellen says. 'She had her sixth birthday in September 2017, and I'd say October was the beginning of her decline. By mid-November she was losing the ability to move, and she died on the 11th of January.'

Shortly before Amity died, Mary-Ellen said they learned that Haber's institute had set up a tumour bank, to allow parents to donate their child's brain tumour so it could be grown in the lab and studied to increase the potential for a future cure. For Mary-Ellen, it was an easy decision. 'We were like, absolutely we will donate the tumour – for a couple of reasons. One, we liked the idea of burying her without the tumour, and Jackson likes the fact that the tumour is probably being tortured in the laboratory. But secondly, what progress is there going to be for other children if we don't do that? Also, Amity said at a very young age that she wanted to be a scientist and so we figured, well, this is her small way of contributing.'

After Amity's death, Mary-Ellen started attending fundraisers for the institute, where she got to know Haber. 'She is pretty much always there at those functions and she is a powerhouse. She can do science but she can also do people – she's a fabulous communicator. I can see why she's been in that role for so long.'

In February, another mother, Kathryn Wakelin, gave a speech at an institute event about losing her eight-year-old son Levi to DIPG, the same disease that took Amity's life. After Levi's death, the Wakelin family helped to set up Levi's Project, a research project run by Haber's institute that has so far raised $2.6 million for DIPG research. 'I spoke [at the event] about our journey as a family through illness and death and bereavement, and how it changes you and your outlook on life,' Wakelin recalls. As she was speaking, she noticed that Haber, who was in the audience, was crying. 'It was amazing,' Wakelin recalls. 'She had tears in her eyes and it was not the first time she had heard our story. That's Michelle for you. She has genuine compassion. It's not just her job but her mission. You can feel it. And it makes parents like us want to be a part of that mission.'

Now Haber and her institute are embarking on their most ambitious move yet. They want to expand their successful Zero program beyond children with high-risk cancers to include all children in Australia with any type of cancer. So for the first time, every child diagnosed with cancer in Australia will have the chance to access the personalised medicine that Haber and her team have designed. That will occur by the end of this year, and in 2025 the institute's new home, the country's first Children's Comprehensive Cancer Centre, is due to open at the Sydney Children's Hospital in Randwick.

Haber insists that she is not being a dreamer when she says that she now believes childhood cancer in Australia will be conquered within a generation. 'I don't see that as aspirational, I see that as reality,' she says. 'Because of medical research we've gone from zero survival to 70 per cent to 80 per cent and 85 per cent. One day we will get to zero children dying of cancer and there will be drugs available that will make this a chronic, manageable disease,' she says.

She breaks into a smile and fixes her eyes on me. 'That is the future.'

✳ *This little theory went to market*, p. **135**
Doing drugs differently: For public health, not profit, p. **162**

DOG PEOPLE: HOW OUR PETS REMIND US WHO WE REALLY ARE

Amanda Niehaus

As I write this, my dogs are curled up in the sun on the bench beside me. It's winter in Brisbane, bright and cool, and the smaller of the two, Quito, is wearing a cable-knit jumper, startlingly pink like fairy floss. It has a turtleneck of sorts, and little holes for his front legs, and when I put it on him, it triggers a 'cosy' reaction, and he instantly runs to the nearest fluffy location and snuggles in. Today, he's chosen a thick, round pillow on top of the thinner cushion topping the bench. The larger of my two dogs, Tukee, can't seem to move when he's got a jumper on, so he's using Quito as a hot water bottle. I don't blame him. I can't quite feel the tips of my fingers, all too naked, as I type.

My husband is overseas, my daughter back at school. In the temporary quiet, I'm thinking about my father, who died not so long ago, and about my mother and sister, who are adjusting to a new, closer way of living, 13 000 kilometres away from here. I could text them, or FaceTime, or scroll through a social feed – but I find it hard to feel connected over such a distance. I am too conscious of my own small face in the corner of the screen, the garble of my comments online. I am not quite my real self.

We're social animals, humans – from the wiring of our brains to the shape of our societies. If recent pandemic lockdowns taught us one thing, it's that we need to be physically close to each other, to socialise not just as avatars or gigabits but as live, warm, fallible bodies. Our dogs knew this ages ago. Have you ever tried to say hi to your dogs over Zoom? Most of the time when I try, they don't even acknowledge my voice. I'm an in-person person, and they are, too.

Every now and then, our resident juvenile brush turkey, who we call Dr Alan Grant, comes to the sliding-glass doors begging for pieces of apple, and when she does, the dogs leap up to defend the house. It's a fair call, actually – Dr Grant would very much love to be inside, as she was once, accidentally. She got a taste for it in those forty seconds: the big grey couch, the bookshelves, Netflix. This is where all apples come from.

But this isn't a story about Dr Alan Grant – it's a story about my dogs, and all our dogs, and what they tell us about our own humanity. How, tens of thousands of years ago in a half-frozen world, humans invited wolves into their family circles and domesticated them, and in doing so made sure that today we never really have to be alone, even when the internet drops out.

The Pleistocene was cold, made colder by a mis-tilt of the Earth 115 000 years ago that caused ice to form across what is now Canada, the US and Europe. Where there wasn't ice, the Northern Hemisphere was predominantly wide, grassy plains on which giant species evolved – mammoths and mastodons, sabre-toothed cats, one-tonne ground sloths, dire wolves. By the time the Ice Age hit, modern humans, *Homo sapiens*, had been evolving in Africa for 150 000 years; some of these humans moved not away from the ice but into it. By the time the Ice Age ended, more than 100 000 years after it began, humans had colonised not only Europe and Asia but Australasia and the Americas, and they took dogs with them.

Grey wolves in the Pleistocene, unlike their larger dire cousins, were around the same size as grey wolves now – and for most of our shared history, humans and wolves have lived as competitors, each a threat to the other's survival. But in some ways, our two species were made to connect. Both are intensely social, live in family groups and communicate using a wide range of facial, bodily and vocal signals, often subtle; we reconcile after fights, play as adults and use touch to build and maintain social bonds. Wolves use sounds, postures and expressions to reinforce group dynamics, including dominance or submission, and to express pleasure, warn intruders away or rally others. Like humans, they have an acute social awareness that enables them to communicate their own intentions effectively, understand the intentions of others and modify their behaviour appropriately in response.

Precisely how, when and where dogs were domesticated remains a mystery, but recent DNA studies suggest that at some stage, as long as 40 000 years ago, humans in north-eastern Siberia, as well as possibly eastern or central Eurasia, began to accept wolves into their social groups. So began the first domestication of an animal species, the grey wolf. At first this process was unconscious – a tenuous social alliance – but eventually humans bred the animals for their speed or fur or other desired traits, producing the 360-plus dog breeds we know and love today.

We didn't plan, at the start, to become chihuahua people. In March 2018, my daughter and I ambushed my husband while he was on a work trip in Brazil. He'd been out with colleagues, came home late and was a little tipsy, and in that state it was easy to convince him over the phone that we needed a second dog.

Our family dog, Lupa, was then thirteen, and a somewhat old thirteen at that, due to a longstanding metabolic issue. Lupa was a rescue, discovered by our friend during creek restoration on Brisbane's northside, and she was on the independent, quiet end of the Jack Russell/fox terrier spectrum – a pottering great-aunt sort of dog from an early age. She was affectionate in a reserved way and indiscriminate in that affection. Her tail never stopped wagging.

But Lupa needed a sibling. She was slowing down, and a puppy would liven her, we argued, keep her from being bored. We were crafty and compelling. A few days after Robbie returned from overseas, we went to meet a litter of puppies.

Whippets are among the most mild-mannered of all dog breeds – as adults, anyway. The whippet puppies were plump and sleek like fish, and boisterous, and as we tugged ropes and threw plastic bottles for them, the adults, refined and languorous, watched on like marble gods. But our small townhouse, small garden would not fit this kind of art – and we felt that we needed a dog that was shorter than Lupa, and lighter – someone she could boss around if she needed to. A bigger dog, we worried, would be hard for her to handle.

So we got a chihuahua.

We knew very little about chihuahuas at the beginning, aside from what the American Kennel Club website told us: 'A graceful, alert, swift-moving compact little dog with saucy expression' and 'attitudes of self-importance, confidence, self-reliance'.

And what our friends said, without words, when they raised their eyebrows at us.

Domestication is a long, messy process, involving many generations of animals who are selected and bred for particular features. We can see in our various dog breeds today what some of these traits might have been: protection, hunting, warmth, companionship. In 1959, Russian biologist Dmitri Belyaev began a groundbreaking experiment with foxes to demonstrate how the domestication of wolves to dogs might have occurred, based only on variation in behaviour among individual animals.

Foxes coexisted with wolves and humans in the Pleistocene, but because they are smaller and more solitary, they were more likely to be eaten than tamed. Yet these same characteristics made them ideal for scientific study. Many generations of foxes could be easily and quickly raised and bred in captivity – and, in fact, they already were. Belyaev began his experiments using foxes reared commercially for their fur, which were as aggressive and averse to humans as any wild fox would be. However, there remained some variation in the aggressiveness or docility of individual foxes. The scientists selected the most tame or docile individuals in each annual litter and bred them with each other – then bred the most docile of those pups with each other, and so on. Researchers developed relationships with the foxes, cuddling and feeding them. By the fourth generation/year, some fox pups were wagging their tails when humans approached them, as dogs do; by the sixth generation/year, around 2 per cent of the fox pups not only wagged their tails when humans arrived, but whined when humans left, whimpered and licked their human carers, and chose to be around people. By the thirtieth generation, almost half the fox pups behaved this way.

Although the scientists selected animals for their temperament, the foxes' physiology and bodies also evolved. In only fifteen generations, the domesticated foxes' stress hormones were half the levels of their wild counterparts, and more recent research has shown key changes in serotonin metabolism in the brains of domesticated foxes, in ways likely to improve mood stability and 'happiness'. Domesticated foxes developed bicolour faces like border collies or white star markings on their heads, curly tails or floppy ears. Overall, their faces became more juvenilised – simply said, as they were domesticated, the foxes became cuter.

I was drawn to Tukee from his online photo, blurry as it was, and tried to go and meet him with an open mind. He had lived on a farm for his first six months, and when his owner let him out of the house he sprinted around all the sheep in the front garden, stopping and starting, playing his own strange game as if to charm us. He had a thick, off-white mane around his neck and chest that, against the shorter blond of his coat, seemed like a ruffled shirt, a nod to royal or rockstar tastes. He was interesting and unpredictable in the right sort of way: it was love at first sight.

We brought Lupa out from the car, and she completely ignored him.

Tukee immediately accepted us as his people, including Lupa, whom he tracked as she moved among patches of sun. Lupa spent much of her life stretched out toasting luxuriously, solitarily, on the floor, stairs or deck. Tukee spent much of that first year cuddling her as she did so. She didn't love it, but he more than compensated by teaching her to bark again. Lupa, almost deaf for two years, had stopped telling us when she was hungry or when the possums were out. When he was hungry and being ignored, Tukee learnt that he could yap in a high-pitched tone that Lupa could hear, and elicit loud, annoying barks from her that we responded to. Lupa got her bark back.

When Lupa died eighteen months later, I held my hand against her head as her eyes went empty; I needed to touch her as she left as I had touched her in her living. We had a family memorial for her; I eulogised her online. Tukee stopped eating.

In late 2021, my father's cancer advanced suddenly, and after a few weeks in ICU he was brought home in an ambulance to finish dying in the living room of my childhood home in Iowa. The hospital bed was erected between the two lounges, facing the small television and the clock on the wall. That coming-home day was the best day he had in all the days I saw him before he was gone – he was home, with his people. He couldn't stop smiling.

I have failed again and again to write about those weeks, as he became less of his mind and more of his body, as I shifted his bony frame from side to side, cleaned his mouth, chest, anus. I thought there was time to revisit the family history, the stories I'd forgotten, but it was already too late. He was so tired, dying, and my mother, warm and alive, curled up in her chair and ached.

I have tried to understand what those three months meant, more than just everything. On the day he arrived home, I posted a picture of the ambulance on Instagram, captioned 'Dad's home! ♥☆', as if that captured even a fraction of what I wanted, or needed, to say:

We die, we are dying, there is hope, I'm alive, I can do this.

How will I do this?

I needed to share the experience with other humans; I needed to connect. But it came out all wrong, too artificial, unconsidered – a snapshot when I needed an essay. From all over the world, my friends, people who loved me or understood what this time was, responded with kind words, love hearts. I sat alone on the couch, apart from my virtual self.

Humans probably brought dogs with them when they migrated across the frozen Bering Strait from Siberia into the Americas 15 000 to 33 000 years ago. Prior to that migration, there were no humans in the Americas; there were wolves and coyotes, but – without our early influence – they did not evolve into dogs. When the Bering Ice Bridge melted at the end of the Ice Age, humans and dogs in the Americas were isolated from the rest of the world for at least 9000 years, until the Europeans arrived in the late 1400s and early 1500s.

In seasonally colder regions of North America, Indigenous peoples kept larger, husky-like dogs that improved hunting success or produced warm, woolly fur. Further south, in what would later become Mexico, the Colima, Maya and later Toltec and Aztec peoples bred small mute (*techichi*) or hairless (*xoloitzcuintli*) dogs, which were sometimes used as food and were buried with people to guide them to the afterlife. For a long time, it was thought that the Aztecs' small dogs were lost when they bred extensively with European dogs after colonisation. But descendants of the *techichi* were rediscovered in the 1800s in the Mexican state of Chihuahua, and quickly became popular internationally.

Today, only a few American breeds still show evidence of their precolonial ancestry: the modern, still-hairless *xoloitzcuintli* retains 3 per cent of its ancient DNA and the *techichi* persists as 4 per cent of the Mexican chihuahua, no longer even remotely mute.

Quito came into our lives in part to help Tukee through the grieving process. He is two years younger and significantly smaller, almost half the size. He's what they call a tricolour, mostly black but with ruddy eyebrows and a white crest on his chest like a superhero. He loves being carried, gets the hiccups, is shy. He barks or growls at the postie, at the crows and, on occasion, at Tukee when he tries to cuddle him, and when he barks his hackles rise – making him look, suddenly, like a tiny black wolf.

Quito is a simple creature, motivated more by food and temperature than affection. He is shaped like a piglet and gulps his kibble so quickly and voraciously that his farts are horrific. We've invested in every kind of bowl, mat and puzzle game you can imagine to slow him down. If we were all trapped together somewhere, and food was on the outside, he'd adeptly catch flies for us to survive on – or he'd eat us, guiltily, until he exploded.

As far as we know, our dogs are entirely unrelated to each other, yet they behave as siblings – older and younger – and we find ourselves treating them that way, too. Quito gets concessions for being the baby; we expect more from Tukee as elder. Several times a day, they play and they fight, and I think of my little sister, our long-ago games with her Lego, He-Man and Orko, my Barbies hidden in her closet.

On Instagram you'll find more than 35 million images tagged *#chihuahua*. Among them are chihuahuas in prams or wearing cute outfits or wagging their tails so hard they might just take wing – but all of them feature a small dog with slightly bulgy eyes and ears like ship sails. For all the diversity among individual chihuahuas, there is still an energy that connects them, that connects us as their owners across more than a thousand years.

In fact, google 'colima dog' and you'll see the same dogs as crafted by artists in pre-Colombian Mexico, ranging between 2200 and 1400 years old. In the collection of the Los Angeles County Museum of Art, a 42-centimetre ceramic colima dog vessel lies on its belly, grasps a corn cob between its forepaws and gnaws it, pointed ears up, tail up, hind legs extended behind like chicken thighs. The clay is blond, smooth with age. Give Tukee a stick, and this is how he chews it.

At Mexico's National Museum of Anthropology, a palm-sized red clay dog barks or snarls – ears alert, tail curled over his back, belly round. Quito's belly is the same, despite the care I take with his food, because it's his shape: his torso is short, his ribcage broad, and he will always look delicious. You will see 'Reclining Dog' at the Metropolitan Museum of Art in New York City, the colour of a burnt-sienna crayon, lying attentively, as though his master has asked him to. His forearm wrists are curled in together at the front, in that way that seems so uncomfortable when my own dogs do it, which is when they're most relaxed.

And my favourite, also at the National Museum of Anthropology in Mexico, which is not the most beautiful but simply speaks the loudest: a small ceramic dog wearing a human mask, with dark brows and hawkish nose. The clay is red, shiny – mottled with black, like age spots. According to experts, the mask signifies that this colima dog was supernatural. But to me, the mask says something else, too – about our long history with these animals, and the relationships we have developed with them.

Here is how you meet my dogs: When you arrive, there will be barking. Do not make eye contact; do not make abrupt movements or gestures; do not try to pet them or acknowledge them in any way; do not push them away when they sniff your legs, groin and face assertively. Sit down, and they will need to start again. Stand up, and they will need to start again. Go to the bathroom, emerge (to them) as a completely new person, and start again.

After a year or more of meeting you repeatedly, they will recognise you as part of our extended pack, as family. Then, everything changes. You arrive, and when you have at last patted them hello, they will tremble, kiss you profusely, and you will have warm cuddles in your lap, your arms, over your shoulders whenever you might like a silky rush of oxytocin – when you don't even know that you need it. There is something very special about cuddling an animal the size of a baby human when there are no baby humans around.

I can't yet define the gap my dad left – it's more than his life, our L-shaped house, the fruit trees he planted, the habit he had of collecting. More than my memories of forty-six years, what decades of photographs say, his voice on my sister's phone, his words in her inbox. I don't want to worry I wasn't enough. What matters most were not the words or the images, but the hugs, the hands, how he carried me on his shoulders, how I tried to hold him up when he fell.

I came home to Brisbane from those last months in Iowa, winter to summer, stepped out of the car and through the gate, and my dogs were there. I needed them first, the squeals and spins, how they paw the air like circus horses, how they kiss without humility or restraint, how they wear their small anxious wolf-pack hearts on their sleeves, as I do, as it pains me to do. My chihuahuas remind me that we're not all one thing, that we're vulnerable, reactive; we need each other.

Humans are always reading the minds of other humans. We are sensitive to countless overt and subtle cues – body language, vocal tones, facial expressions – and we use these to predict other people's thoughts or emotions and to adjust our own behaviour accordingly. We draw on these same abilities when we read or tell stories. We use written cues to transport into fictional characters' bodies, and we understand even as young children how to read an audience and respond in ways that keep people listening. Humans evolved the means to both embody others and respond to others' bodies, but did so before letters or phones or computers, when we still lived in small family groups. Before likes and emojis, re-shares and hashtags, we were in-person people, all of us.

Our dogs perceive our feelings and understand our gestures better even than our nearest relatives – the apes – do. Wolves avoid eye contact, as it's a sign of aggression, but even as puppies, dogs focus on human faces, seek out our gaze. My dogs don't care what words I choose online, who I seem to be on Facebook. They understand their names, 'bedtime', 'dinner', 'upstairs' and 'inside'; beyond that, what I say doesn't matter – it's the tone of my voice, the tension in my body, and what I *do* that counts.

I point, and they look past the end of my finger; I look at them, and they know who I am.

✱   *Can a wild animal make your house feel like home?*, p. **51**
Call of the liar, p. **198**

CAN A WILD ANIMAL MAKE YOUR HOUSE FEEL LIKE A HOME?

Tabitha Carvan

Every evening in our old house, always around the same time, we'd hear the rumpeta-rumpeta-rumpeta on the roof.

'The possums are up,' someone would announce, because this was only the beginning of a whole nighttime of noises.

There'd soon be the rustling bushes, the overladen branches scraping on the windows, that unreal gurgling sound, the intermittent screech. When the ABC ran a poll to find Australia's favourite animal sound, the 'unsettling' vocals of the brushtail possum came in third-last and I wondered how they did so well.

In the daytime, the possums snoozed in our garage, curled up inside the roller door, or in between the rafters, tucked into a comfy nest of branches they'd dragged in from the bay tree. We'd try to close off their entrances, and they'd just find a new way. They stank, but they also had big brown eyes and sometimes adorable babies clinging to their backs.

When we moved out, the new tenants told us that the landlord had taken the garage off the lease. They could use it if they wanted, but it would be at their own risk. The garage belonged to the possums now. I had to laugh. This had always been the case, but now it was official.

Our new house is only one suburb away but it's almost entirely possum-free. You can leave an empty cardboard box in the garage and the next day it remains an empty cardboard box, without even a hint of possum inside. When you put the bins out late, you'll see possums dashing up the trunks of the trees on the street, but they must have no interest in using our roof as a thoroughfare, because we never hear them.

We sleep soundly, all through the night.

But something is missing.

Having complained about the possums for so many years, it's ridiculous to even think it, but it's true: I miss them.

Fifteen years ago, an urban geographer from Macquarie University, Associate Professor Emma Power, interviewed a group of Sydneysiders about their experiences, similar to mine, of cohabiting with possums.

Her findings showed that participants had 'paradoxical' feelings about their housemates. Yes, the possums were destructive and disruptive, and yet also, they were a comforting presence in the lives of the residents.

They said that while they wouldn't necessarily choose to cohabit with possums, the fact they did made them feel more connected to their surroundings. They gave the possums names and took pride in them, like they were family members, just ones who happened to use the ceiling cavity as a toilet.

The possums, Associate Professor Power concluded, created a feeling of 'homeyness'.

I read this study and I knew what my problem was. I didn't exactly want possums in our new house, but without them, it just didn't feel as homey.

I started asking people if they knew what I was talking about. In response I heard a stream of stories: there's the garden magpie called Maggie; Old Seven Legs, the huntsman, who lives in the spare room; the skink who always stops to say hello at the washing line.

My colleague Elsie told me about the time her family received a postcard from a friend addressed to all human members of the household and additionally: Dotty (the dog), Shadow (the cat), Bluey (the blue-tongue lizard that lived in the backyard), and Penelope (the possum that lived in the roof). Upon reading it, Elsie's mum responded, outraged, 'What about Bocky?' Bocky was the frog that lived in the garden. All these creatures lived at the same address, after all.

But what business does a wild animal have in contributing to our domestic bliss? And isn't the whole idea of a home that it keeps the wild things, and their unsettling noises, on the other side of the walls?

Dr Pele Cannon tells me, fondly, about the wombat who used to turn up at the window of her family's home every evening to sneak a look at the humans inside.

She researches human and non-human coexistence at the Australian National University. In the acknowledgments of her PhD thesis, she thanks 'all the non-humans throughout my life who told me their stories, taught me their ways, and didn't let me get away with anything'. She seems like the right person to ask about the homeyness generated by wild animals, and she is.

'I love this question,' she says. 'And I think the answer is it's about feeling grounded.'

As humans, we can get very up in our own heads, she explains. 'But when you see an animal doing its animal thing, especially over repeat visits, it really brings you back down to Earth, and into our physical, embodied existence.'

Anyone who has looked up from their laptop to momentarily watch a bird outside their window knows this feeling. There's a disjunction, between the supposedly important contents of our computer screen and the bird, outside, unbothered by them, which makes us take a reality check.

'Our homes are so curated and organised, that another being – especially a wild, uninvited one – disrupts that order,' she continues. But it's a good kind of disruption: 'It forces us into the presence of the rhythms of life. It's humbling.' (Yes, I think. Like our garage was humbled.)

But if the animals are just going about their animal lives, isn't it actually a case of Main Character Syndrome to imagine there's any connection going on between us? Presumably they don't care about us at all.

Dr Cannon spent two years living and studying in a wolf sanctuary in the United States. She says it taught her that actually, animals perceive us well before we perceive them, and from far greater distances. There are so many more birds, she says, watching us potter around in our garden and walk to the bus every day than we could ever realise.

'We are a component of their world,' she says. 'And when we notice that, it contributes to the homeyness. It's a participatory feeling with nature, and that's quite soothing.'

And I have to accept that this is actually right, even though it feels entirely against the odds. The possums pooing on our deck really were soothing.

But it shouldn't be that surprising. As Ruby Ekkel points out to me, this idea of a wild animal offering comfort isn't new at all.

Ekkel is doing a PhD at the Australian National University on European colonisers' changing interactions with Australian native animals. She notes that when homesick settlers deliberately introduced animals – invasive species, as we now know them – they were doing exactly this: trying to 'over-engineer the feeling of homeyness'.

It's a story most of us know as an ecological disaster, but it's also one which proves 'that wild animals, not just pets, can be used to try to foster the pleasures and familiarities of home'.

Around the time of Federation, Ekkel continues, the settlers' attitudes towards native species changed. The more 'at home' they felt on colonised Aboriginal lands, the more connected they became to Australian native animals as their 'fellow citizens'. But this connection wasn't extended to all native species equally. Not all animals are sssssoothing.

Associate Professor Gavin Smith is a sociology lecturer and also a snake ecologist, but he prefers the description 'snake intermediary'.

For the past four years, he has managed the Canberra Snake Tracking Project, which monitors the behaviour of eastern browns once they have been removed from people's properties and translocated to the bush.

He agrees that we experience some animals as homey, but what's more interesting is that we don't experience them all that way.

'We put animals into a hierarchy of desirability,' he says. 'And, obviously, my work over the years has been with an animal that sits at the very bottom of that list.'

When we see an eastern rosella on the lawn, he says, it reminds us that our home is part of a diverse and beautiful ecosystem; we feel lucky. But when a snake slithers past our back shed, we now experience that space as our 'private property', and the snake as a trespasser who doesn't belong.

This shows a 'profound misunderstanding', Associate Professor Smith says, of 'the animals which inhabit this country, and which have done so for millions of years.'

'It's humans who have encroached on, and fragmented, the habitats of snakes and other wildlife,' he says. 'Snakes are willing to share resources with us, if we give them some room. But this is a practice that humans and their companion animals are often reluctant to do, sadly.

'I don't think most Australians have really connected deeply and richly enough with the diversity of nature surrounding them. Many have detached themselves from their natural environment. Sometimes we do have these little epiphany moments' – like when we see a lovely bird at the window – 'but nature is not something we can just switch on and switch off as we choose. It is something we are absolutely, intrinsically, connected with all the time.'

The animals aren't visiting us in our homes, he says. We live in theirs.

That participatory feeling which Dr Cannon mentioned – it's us participating in their world, not the other way around.

'The wombat coming to the window doesn't connect me to nature,' Dr Cannon says. 'It reminds me that I am connected already. You don't need to get out into the wilderness to have that connection. You just need to appreciate the things that are around you all the time.'

'Let's get one thing straight. A house is not the same as a home,' writes the essayist Meghan Daum on the subject of architecture. A house is a physical entity. 'Home is an idea, a social construct, a story we tell ourselves about who we are and who and what we want closest in our midst.'

It only seems strange that wild animals can change how we feel about our home if we see them as two separate things: the outside and the inside. But actually, homeyness seems to come from the blurring of those boundaries. It's about rescinding control and allowing your surroundings, and its other occupants, to have more of a say. And it was that feeling, not the possums in particular, which was missing from my new house.

Some animals make it impossible for us not to listen to them, but we actually have lots of other housemates, just ones which can't get our attention in the same way. When three Brisbane researchers embarked on a lockdown project to catalogue every animal living with them in their Queenslander and its garden, they identified 1150 different species. It shows that every house has the capacity to be homey, if we let it.

'You grow a home,' Daum continues. 'You let it unfold on its own terms. You wait for it.'

It's 4 am and I am lying awake in our new house, which isn't that new anymore. Also, I am no longer sleeping through the night. The extremely loud, overdramatic call of the male eastern koel – Australia's eighth least-favourite sound – is already going strong. It's the one that goes *woo-woooo-WOOOO-WOOOOOO*. I'll admit I have cursed this bird in my half-sleep, many times.

It turns up in the spring to find a mate, which is hard because as a type of cuckoo, it was raised by a different species of bird. The reason its call is so loud and carries so far is because this is the only way it can find another unfamiliar koel.

It's a good reason. And now I know I would miss it if it wasn't there. It's also still a bit annoying.

'I'm listening,' I say into the darkness.

And he answers back. '*Woo-woooo-WOOOO-WOOOOOO*.'

For the both of us, it sounds like home.

✱ *Western Australia had its hottest summer ever, but climate change barely made the news*, p. **22**
The consciousness question in the age of AI, p. **76**

OF MOTHS AND MARSUPIALS

Kate Evans

Linda Broome pulls herself up a near-vertical slab of granite, leaps nimbly over a snow-lined fissure, and dives head-first into a crack in the rocks. At 67, she knows Mount Blue Cow like the back of her gloved hands. Every November for the past three decades, she has lead volunteers up this 2000-metre peak, and a handful of other high-altitude boulder fields in Australia's Snowy Mountains, where the states of New South Wales and Victoria meet. The team visits to monitor the population of critically endangered *Burramys parvus* – better known as mountain pygmy possums.

November is springtime in the high country. Drifts left by a late snowstorm are melting, and frogs croak enthusiastically in the fens. Anemone buttercups – rescued from near extinction 65 years ago when the government banned cattle grazing here in Kosciuszko National Park – wave white-and-yellow among snowgrass and snow gums and fragrant alpine mint, while little ravens caw from the windswept sky.

The nocturnal possums have hibernated under the snow all winter, dropping their body temperature to just above freezing for up to seven months. Then, one night a few weeks ago, they woke up. Broome, a threatened-species officer at the New South Wales state environment department, retrieves a rectangular aluminium trap from the crack and backs out. She carefully reaches inside, past a lining of cushion-stuffing, and pulls out a whiskery, mouse-like creature with black eyes and brown-and-gold fur.

She holds it by the base of its long, scaly tail, which curls around her finger. 'It's alright, darling,' she coos. 'You're so sweet.' Broome and volunteer Carlie Armstrong work together to insert a minuscule microchip into the folds of skin at the possum's neck, clip a snowflake-sized disc of skin from its ear for genetic analysis, and check for parasites.

They place the creature in a cloth bag and weigh it: 34 grams. This one is female, with a handful of jellybean-like joeys in her pouch, and seems unconcerned as she perches on Broome's finger and sips water from a bottle lid. While Armstrong uses the metal trap pin to nudge the possum's droppings into a vial, Broome releases the animal. She scurries over the neon lichen and disappears into a crevice.

The little marsupial's dual mission for the coming summer is to raise her young while doubling her body weight, so she's fat enough to hibernate again. Fortunately, just as she and the other possums were stirring beneath the boulders, breakfast arrived on their doorstep.

Every spring, vast numbers of bogong moths (*Agrotis infusa*) migrate as much as 1000 kilometres (600 miles) from the western plains of New South Wales and Queensland to the high country, crawling into caves and among the rocks to avoid the heat – a summertime form of hibernation called aestivation. The alpine possums eat other invertebrates, as well as fruits, seeds and nectar, but the fatty bogong moths are their favourite food; at the highest elevations, the migrating moths make up as much as 50 per cent of their diet. 'It's a massive movement of protein into the mountains,' says Broome – one that feeds not just possums, but ravens, lizards and other small mammals.

In the past few years, though, New South Wales and the rest of south-eastern Australia have ricocheted from drought, to fire, to flood, threatening to unravel the ancient, intertwined relationship between the moths and all the animals that depend on them –

including the few thousand mountain pygmy possums known to exist.

From 2017 to 2020, prolonged droughts in Queensland and western New South Wales caused moth numbers to drop from more than 4 billion to perhaps 20 million. As a result, baby possums starved in their mothers' pouches. Meanwhile, bushfires ravage other mountain food sources, and Australia's snow is itself endangered.

The dance between moth and possum demonstrates the complexity of global climate change, and how extreme weather in one place can alter environments thousands of kilometres away. Together, these creatures have become symbols of what Australia stands to lose in this warming century: unique species, ecological relationships, and even entire ecosystems, alongside human homes, livelihoods and lives.

But despite their small size, low numbers, and harsh habitat these possums are surprisingly resilient. The fossil record suggests that the tiny marsupials have adapted to dramatically altered climates in the past, and they may do so again – provided they have some help.

Until the 1960s, scientists considered the mountain pygmy possum nothing more than an interesting fossil. In 1894, Scottish doctor and palaeontologist Robert Broom had found minute Pleistocene-age teeth and jawbones in a limestone deposit near the Wombeyan Caves, between Canberra and Sydney.

The jaw featured a most unusual tooth, 'unlike that of any known marsupial,' Broom wrote – a large premolar with six deep grooves that gave it a serrated edge. Speculating that it belonged to a missing link between kangaroos and tree possums, he gave the presumably extinct creature its scientific name, borrowing *Burra* from a local Aboriginal word for the rocky caves, *mys* from the Greek for mouse, and *parvus* from Latin, meaning little. 'So it was a small rock mouse, which was very prophetic,' says Linda Broome

(no relation to Broom, but she appreciates the resonance: 'It's destiny that I studied this thing.')

A few more fossil remains turned up over the following half-century, and in the 1950s, David Ride, director of the Western Australian Museum, used vinegar to dissolve the rock encasing the fragile Burramys bones. With this clearer view, he decided they were more closely related to Australasian possums – *Phalangeriformes* – than to kangaroos, and that the Burramys puzzle had been solved. (American opossums are also marsupials, but are more distantly related.)

Then, in August 1966, people staying at a Mount Hotham ski lodge in Victoria noticed some strange furry creatures stealing bacon from the stove. Australian naturalist Norman Wakefield took a look at a captured one and recognised its distinctive ridged tooth immediately. 'Burramys had come to life,' Ride wrote of the event. 'The dream dreamed by every palaeontologist had come true. The dry bones of the fossil had come together and were covered with sinews, flesh and skin.'

Because other species of pygmy possum live in trees, ecologists initially assumed that the possums had hitchhiked to the lodge in a load of firewood. But by the early 1970s, they had been captured in the wild at various alpine locations in Victoria and New South Wales – even, in 1972, among the boulders near the treeless summit of 2228-metre Mount Kosciuszko, Australia's highest peak. A small rock mouse, indeed.

In the early 1980s, as developers were building a new ski resort at Mount Blue Cow, a national parks contractor discovered a population of pygmy possums precisely where the runs were being laid out. The state government set out to find someone to assess the resort's impact on the animals.

At the time, Linda Broome was living in Logan, Utah, USA, completing her doctorate at Utah State University. Having just spent four years tracking deer mice across the rolling sagebrush country of Wyoming, she was probably the only person from

Australia – that mostly sunburnt country – who had worked with small mammals in the snow. She got the job, and came home. When she first arrived at the foot of Blue Cow in 1986, she looked up at its steep, rocky profile and thought, 'Hell's teeth! What have I done now?'

For the next few years, she trapped the possums and fitted them with wedding-ring sized radio transmitter collars she had soldered together out of cable ties, dental acrylic and hearing-aid batteries. She tracked their habits and movements, mostly at night. Once, she dodged a dynamite blast while setting possum traps in a boulder field that workers were clearing for a ski run.

And as winter blizzards raged, she clambered over the mountain in snowshoes with her radio receiver, finding the possums unexpectedly still. That was how she discovered that they hibernate – a rare behaviour for Australian mammals.

Over the decades, she and other researchers identified three genetically distinct populations: the one here in Kosciuszko National Park, another at Victoria's Mount Hotham and the Bogong High Plains, and another at Mount Buller, also in Victoria. They have watched the possums weather fire and ice, dry years and wet ones. The animals' total numbers have fluctuated between around 2000 and 3000 adults, depending on climate, food availability and predators – and on the mysterious peregrinations of the bogong moths.

In September 2000, a strange cloud appeared on satellite maps, approaching Sydney. Weather forecasters worried that rain would disrupt the Olympic Games Closing Ceremony. But the 'cloud' turned out to be a huge swarm of bogong moths. Attracted by the stadium lights, they tumbled dizzily above the spectators, and threw themselves against the soloist as she sang the Olympic hymn.

It was not the first time that bogong moths had been part of human celebrations. For the Aboriginal peoples of the high country, bogongs have been an important seasonal marker for thousands of years, says Jakelin Troy, a linguist and the Director of Aboriginal and Torres Strait Islander Research at the University of Sydney. Troy's people are the Ngarigu, and the Snowy Mountains are her Country. That word, Country, means more than just land. It encapsulates Aboriginal peoples' embodied experiences of being immersed in the environment of their homelands – and their duty to protect them.

'Bogongs are just such a core part of our story,' Troy says. 'They blend in so beautifully with Country. They look like the granite, they look like fallen timber – and they have been this important source of food for my people.' They're also delicious, she says. 'Get all the mothy stuff off and they're like peanut butter.'

The moths feature in stories and songs across south-eastern Australia. At Uriarra Station, half an hour's drive from Australia's capital, Canberra, Paul Girrawah House of the Ngambri people introduces me to a sacred site. The Uriarra Moth Stone is a flat expanse of granite the size of a basketball court, overlooking undulating fields where sheep graze beneath forested hills. 'Uriarra means "running to the feast",' House explains. 'As soon as the bogong arrived, it was the sign. The message went out, and people came.'

His ancestors collected the moths from nearby mountaintop caves and carried them here in dillybags woven from stringybark and reeds. The rock, House says, 'is a feasting table', and everyone, friend or enemy, was invited. Here and in the Snowy Mountains, bogong moths brought people together, fuelling summer festivals where marriages were arranged, ceremonies performed, trade conducted and disputes resolved.

Settler Eliza McDonald, who lived at Uriarra Station in the mid- to late-1800s, recalled watching local Aboriginal women set fires on the stone to heat it and cook the moths. Some years were so rich that it took the people who had gathered weeks to devour all of the insects. It was 'something so far better than [lowland] possum and yams,' McDonald remembered, 'that the ebony skins of the eaters literally shone, and their bodies showed a plumpness quite in contrast with the leanness of normal times.'

By the mid-20th century, colonisation and land dispossession had largely ended these traditions. But the moths remain enormously significant, House says. Today, he has brought his son, Reuben, to see the stone for the first time, and to sing some of the old songs.

'To come back here, there's a feeling of empowerment,' Reuben tells me. The 26-year-old has the image of his Ngambri great-great-great-grandfather, Henry 'Black Harry' Williams, tattooed on his shin. The design is based on a 1901 photograph taken right here on Uriarra Station, where Williams worked as a stockman and labourer. Above Reuben's knee, there's another tattoo: a bogong moth, flapping across his thigh. It's a symbol of his heritage, and of resilience, he says. Persistence in the face of change. The bogong moth 'makes me feel strong,' he says, 'and more proud to identify as a Ngambri man, from here.'

The bogong moth's annual migration is one of the world's natural wonders. Only the epic journey of the monarch butterfly across North America compares, says entomologist Eric Warrant, an Australian based at Lund University in Sweden.

While the monarch takes four generations to complete its migration, the bogong moth does it in one, starting at various sites on the western plains of New South Wales and Queensland, and returning home after their summer mountain sojourn. 'They've never made this journey before, and they have never had anybody to tell them how to get there, and their parents have been dead for three months. This is a terribly difficult thing to do,' says Warrant.

In 2012, Warrant and his students began a series of experiments to investigate how the moths knew where to go, and how far to fly, and when to stop. Being nocturnal, they couldn't use the sun, and the moon's phases made lunar navigation unreliable. The tests showed the moths relied in part on their magnetic sense for navigation, but – like hikers training a compass needle onto a distant tree – they also appeared to use visual cues.

The question was: which cues? It occurred to Warrant that 'the Milky Way is an enormously strong visual stimulus'. To a moth, it probably looks like 'a lovely stripe of light', brightest in the south, and fading in intensity as it arcs into the north.

At his holiday home near Adaminaby, near the northern end of the Snowy Mountains, Warrant put the idea to the test. He built a new lab from scratch in 2017. No magnetic materials were allowed; the roof was made of corrugated aluminium instead of steel, which contains iron. Inside was an arena where the moths could fly in a kind of magnetic vacuum, so the scientists could study the insects' visual sense in isolation.

Using a computer program devised for planetariums, Warrant's team projected the seasonally and geographically correct night sky above bogong moths they'd captured outside during their autumn migration away from the mountains. Inside the arena, with only the projected sky for guidance, the moths knew exactly where to go.

When the researchers rotated the sky image by 180 degrees, the whole population of bogongs turned and flew in the wrong direction. And when they randomised the stars' positions, the moths were completely disoriented.

'It is totally mind-blowing,' says Warrant. The moths have somehow inherited these star-maps and magnetic signposts from their parents, so they're born not only knowing which way to fly, but they can also correct their course on the wing.

And yet, these extraordinary creatures are suffering. Their numbers decreased after European settlement in the 1800s, then remained stable before declining gradually from about 1980. Researchers aren't sure why, but speculate about changes in climate, as well as land use in the moths' breeding grounds, such as clearing, increased grazing pressure, or pesticide use.

Until recently, however, they still numbered in the billions. In a typical year, the caves at the moths' first stopping-point – Mount Gingera, a 1857-metre peak in the Australian Capital Territory's Brindabella ranges – are 'completely carpeted in moths', says Warrant. Each individual overlaps its wings with those of the moth next to it, a behaviour known as tiling. 'They look like scales of a fish. There are 17 000 moths on every square metre of cave wall – it's an amazing sight.'

But in 2017, the rain stopped falling. In the moths' winter breeding grounds on the plains, few plants grew, starving kangaroos and moth caterpillars alike. The following summer sweltered, and when Warrant went up to the caves on Mount Gingera, he found them bare. 'From millions upon millions of moths, to nothing! That for me was shocking in the extreme.'

Large populations can weather one severe year. But for the next three years, rainfall in the state was the lowest on record. Meteorologists estimated that temperatures on the plains were higher than at any point in the previous 2 million years. By the time the drought broke in 2020, Warrant and other bogong moth researchers calculated that the population had plummeted by 99.5 per cent. In 2021, the bogong moths joined the mountain pygmy possums on the International Union for Conservation of Nature's Red List of Threatened Species. That summer, despite two years of good rainfall, surveys of known aestivation sites in Australia's northern Alps, including Mount Gingera, the Main Range and Mount Blue Cow, found few to no moths.

There was a time when Warrant would have laughed at anyone predicting the bogong moth's extinction. Like 19th century Americans marvelling at the superabundant passenger pigeon, he had thought the moths were invincible. Now, he says, it's clear that 'this apparently incredibly resilient insect is very vulnerable to bad conditions'.

It's 6.30 am, and Linda Broome is checking her moth traps at Charlotte Pass, a few miles south-west of Mount Blue Cow. A bowl-like valley cups a clutch of ski lodges, its steep sides lined with gnarled snow gums and striped with chairlifts and pomas. Every year, Broome monitors the population of pygmy possums that lives among the boulder fields beside the road, and she has long kept an eye on their primary food source.

She lifts a funnel from the mouth of a white bucket perched on top of a rock, and removes torn pieces of newspaper. The moths – attracted by a light placed atop the bucket overnight – whirl manically underneath, the scales rising from their wings like dust motes in the light. 'We've only got a few hundred here,' Broome says. It's a partial recovery from the last few dire years, but nothing like the thousands she typically captured before the drought.

Hungry ravens edge closer. Volunteer Zoe Barber reckons they have 'teenage boy energy', and it's true – they strut about in delinquent gangs of five or six, looking for mischief. Torn-off moth wings litter the lichen, evidence of the conspiracy's last feast. Broome tips out the bucket and the moths scurry for the nearest crack. She shoos them deeper, out of the ravens' reach.

Ecosystems are complicated, and it can be hard to untangle cause and effect. But consecutive years of low moth numbers do harm possums – especially those at the highest elevations, where there's little alternative food. In Victoria, when bogong moth populations plummeted in the summer of 2017–18, researchers found possum after possum with their pouches full of dead babies. In one area, 95 per cent of litters died.

'There were no signs of injury, no bacteria, no illness, no viruses,' says Marissa Parrott, a reproductive biologist from Zoos Victoria. 'They simply starved to death.' The government team managing pygmy possum recovery, including Broome, Parrott and others, concluded this was a conservation emergency that required direct intervention. Zoos Victoria got to work on a long-term, possum-specific supplementary food program and came up with a recipe that mimicked the nutritional qualities of bogong moths, plus the other elements of the animals' natural diet. It included macadamia nuts, coconut oil, mealworms, egg white and vitamins. A commercial company made the gourmet mixture up as a dry powder, and then the researchers baked it into 'bogong bikkies'.

The team aimed to create something the possums would like, but not love – more gym protein bar than sweet treat, Parrott says. Trials with captive possums confirmed that the animals would eat the bikkies only if they couldn't get their bogong moths or other natural foods.

In November 2019, Parrott's team successfully tested their concoction in the wild among the struggling possums in the Victorian boulder fields, using a variety of different home-made feeders. But it wasn't until January 2020 that the bikkies really proved their worth. That month, Parrott got a call she would never forget. It was Linda Broome, and she didn't even say hello. 'It's gone,' Broome said. 'There's nothing left.' Bushfires sweeping across vast areas of Australia's south-east had hit northern Kosciuszko National Park, near Cabramurra. The area's tinder-dry boulder fields were home to a thriving population of mountain pygmy possums that Broome and her team, including PhD students Hayley Bates and Haijing Shi, had discovered in 2010.

Broome knew the possums had likely survived, deep in the damp crevices. But when she visited days after the conflagration, she found the still-smoking hillsides devoid of vegetation and insects for the animals to eat, and no water for them to drink. 'Please tell me your food and your feeder worked?' Broome asked Parrott. It was one of the proudest moments of Parrott's life that she could say yes – that the prototypes had been successful, and that they were ready to deploy.

The Zoos Victoria team sent bags of bogong bickie mix and prototypes of the feeders to Broome, and the volunteers got making and baking. Every week for the next two summers, the National Parks and Wildlife Service Discovery Rangers, aided with baking by local schoolchildren, delivered fresh 'bikkies' to 60 feeders stationed across the burned boulder fields.

By the end of 2022, the animals were thriving without support. 'On one of the sites, almost every trap had possums,' says Bates, now an ecologist at the University of New South Wales. Vegetation was returning only slowly, but other prey like bugs and beetles were already crawling around the boulders. The expensive, labour-intensive experiment had worked – proving that in extreme situations, audacious interventions can stave off disaster for endangered species. Unfortunately, the need for them will only rise.

Bushfires are natural in Australia, but their frequency and intensity are predicted to increase as the climate warms. Alpine ecosystems in particular require a long time to recover, especially from consecutive burns. In 2003, for instance, bushfires burned right over the top of Mount Blue Cow. Twisting, skeletal forms still writhe among the boulders – the bleached bones of mountain plum-pine, another favoured food source for the possums. Broome transplanted seedlings to replace them, but two decades later, even though the recent fires spared Mount Blue Cow, they're only just beginning to take.

Then there's the snow – the emblem of the high country, and the source of the water that feeds the fens and the streams. Snow depth and the number of snow days have been declining in Australia since the 1950s, and climate scientists warn that, by the end of the century, the Snowy Mountains may no longer live up to their name.

'The outlook for the alpine zone as we know it is pretty bleak,' says ecologist Lesley Hughes, an emeritus professor at Sydney's Macquarie University, IPCC report author and director of the Climate Council of Australia. Even before it's gone completely, dwindling snow cover will disturb the possums' winter rest. A thicker layer of snow provides more insulation; without it, the animals' nests get colder, which could wake them from hibernation before moths arrive or seeds are available, Broome says. Snow is also a barrier to predators, and warmer winters allow feral cats and foxes to range more freely and hunt possums more easily. In 2002, Broome asked rangers to start trapping and killing cats at Mount Blue Cow. They caught 30 that first winter.

The mountain pygmy possum is often ranked among the Australian species most vulnerable to climate change: global efforts could still restrain rising temperatures, but the possums' high-altitude home has already begun its transformation. Even if conservationists continue to intervene with supplementary food when necessary – without snow, without moths, ravaged by fire and cats … is its extinction inevitable?

The pygmy possums' deep past offers a glimmer of hope. Palaeontology reveals animals and ecologies in the fourth dimension – time. And some scientists believe that insights from the fossil record can offer surprising, if drastic, solutions to the environmental problems we face today.

I meet Michael Archer at his palaeontology lab at the University of New South Wales in Sydney. Mounted mammal skeletons line the walls, and the benches are piled with academic tomes, jars of paintbrushes, bottles of acetone, fossilised skulls and partially prepared snake vertebra. Today, there is also a live ringtail possum in a box; Archer rescued it from the university cafeteria this morning when he stopped by to grab his coffee.

Now in his seventies, Archer was born in Sydney, but grew up 'among hillbillies' in upstate New York, where he learned to make moonshine and play the banjo. When he was 11, he saw some strange shapes poking out of a shed-sized lump of siltstone not far from his house, and chipped bits off it with a sledgehammer and chisel. Inside, he discovered a trove of 370-million-year-old fossils, including trilobites. He packed them into a suitcase and took the steam train to New York City for identification at the American Museum of Natural History. Like dinosaur-loving 11-year-olds everywhere, Archer was hooked. 'Most kids get over it,' he says. 'Some don't.'

He moved back to Australia in his twenties, and since 1976 has led excavations of the extensive fossil deposits at Riversleigh in remote north-west Queensland. As he shows me around his lab, Archer points out hunks of Riversleigh limestone bathing in vats of acetic acid, the stone gradually dissolving to reveal the bones within. I see the spine of a giant flightless bird erupting from a rock. 'Every one of these is a treasure chest,' he says.

One of those treasures is Burramys. Archer's team has found more than 500 pygmy possum bones in Riversleigh rocks 25 to 12 million years old, mainly jawbones and that distinctive, saw-shaped tooth. The fossils are practically identical to the bones of modern mountain pygmy possums. Yet when these ancient ancestors lived, north-west Queensland was carpeted in lush, lowland rainforest. The marsupials' specially ridged premolar allowed them to scurry around on the ground, cracking open nuts and seeds: filling a niche unoccupied by any other mammal.

It was such a successful strategy that the possums stuck with it, changing barely at all over 25 million years, through numerous wild swings in Earth's climate. Around 12 million years ago, the Riversleigh rainforest dried out, and the possums disappear from the record there. 'The animals that couldn't hack it died. But Burramys is a scrubber,' says Archer. 'It's a survivor.'

Fossil Burramys then show up elsewhere in eastern Australia. Around 2 million years ago, the rainforests spread all the way into today's alpine areas. When the trees retreated, the continent dried out, and, later, settlers cleared the remaining forests for agriculture, Burramys got stranded in the mountains. But by diving into the rockpiles, hibernating over winter and eating whatever food was available, the species held on.

All this led Archer to believe that the possum's comfort zone is actually the rainforest, and that their latent genetic adaptability may allow them to thrive there once again – a theory that he and collaborators are now putting to the test.

At Lithgow in the Blue Mountains, 145 kilometres west

of Sydney, and 950 metres above sea level, 14 mountain pygmy possums scamper in cages at a special breeding centre under the watchful eye of owner Trevor Evans. Evans teamed up with Archer and Hayley Bates to design the breeding facility, which opened in September 2022. It's part of the wider Secret Creek Sanctuary, which Evans manages and which hosts a variety of native animals from tiny feathertail gliders to koalas, emus and Tasmanian devils.

Inside the breeding facility, there are 16 separate pygmy possum enclosures – enough to house 100 animals. Each features nesting tunnels enclosed in constructed rocky outcrops, while wall stickers of snow-covered Kosciuszko attempt to make the possums feel at home. Water trickles down the rock faces, keeping the environment cool. As soon as Evans turned the watering system on, the possums 'started to bonk like mad', says Archer – their mating instincts apparently triggered by the moisture. When I visit in November, Evans pulls out a small drawer in the fake rock wall to show me where the possums nest. To his surprise, all that 'bonking' has already produced results: inside the woven stringybark is a mother possum with three tiny, near-blind joeys – the first evidence of successful breeding at the new facility.

That promising start has continued, with more joeys born in the months since. And there's other, contemporary evidence that the plan might work. At Cabramurra, one community of possums lives at just 1225 metres (4000 feet) above sea level, lower than any other populations. Snow is inconsistent there, suggesting that the pygmy possums may not require it. In fact, Bates' doctoral studies suggest that a key limiting factor for possum populations is a permanent supply of fresh water. Hibernation is hard on the kidneys, and the animals seem to need to drink as soon as they emerge from torpor. In the high country, snow provides both water and insulation against the cold; at these lower elevations, a cool, sheltered rocky creek could become a suitable home for mountain pygmy possums.

At the breeding centre, Evans and a series of Archer and Bates's students will monitor the captive possums over the next few years. They're curious to see whether they will still hibernate, given the slightly warmer temperatures and lack of snow. They'll also begin to introduce them to lowland food sources and other animals. Eventually, the team plans to release them into Secret Creek's fenced sanctuary, and then into sufficiently damp, sheltered spots in the wilds of the Blue Mountains. 'What we need to do is basically knock on their genetic door and say, "Hey, wake up! You have the resilience to adapt to this",' Archer says.

Introducing novel species into ecosystems can be fraught, but Archer is optimistic the little marsupials won't have a negative impact. Their original ground-dwelling, seed-eating rainforest niche remains vacant, he says, like an empty glove missing its 'ghost hand'. To make sure, the scientists will closely monitor ecosystem interactions during the trial releases in fenced areas, and adjust the plan based on their findings.

Broome supports the effort. She also remains sceptical about the possums' prospects in the lowlands. Temperate rainforests also experience drought and bushfire, she points out – 80 per cent of the Blue Mountains World Heritage Area burned during the 'Black Summer' of 2019–20 – 'and they're riddled with feral cats', a threat that didn't exist prior to European colonisation.

But Archer believes conservation in the Anthropocene requires radical thinking – and radical action. 'We've created a situation where we no longer have the luxury of preservation. We have to think about these strange strategies, because there may not be any other solutions. In many cases, that means moving things from what are increasingly unacceptable habitats to places where they could survive.'

One full-moon night in the Snowy Mountains in March 1834, Aboriginal women performed a song ceremony, in a place now called Dalgety. 'Gundji gawalgu yuri,' they sang, in a language that would itself almost be lost in the ensuing century. 'Gaba gumadji gugu.'

Watching the women sing and beat skin drums that early autumn night was a travelling European botanist named Johann Lhotsky. He noted down the melody, and likely asked for help to record the words. Several weeks later, in Sydney, he worked with three 'musical gentlemen' to write up sheet music, arranging it for voice and pianoforte in the English parlour ballad style. This 'Song of the Women of the Menero Tribe' – Menero, now spelled Monaro, is the plateau to the east of the Snowy Mountains – is the first known piece of music ever printed in Australia.

Aboriginal Australians have been singing songs of the high country for millennia – its snows and its possums, its caves cloaked with moths. In their ancient cultures, story and song have the literal power of creation; songlines trace the movements of ancestral beings as they made the landscape, and encode navigational information and traditional knowledge about animals and plants.

Those singing women might have been Jakelin Troy's people, her own Ngarigu ancestors possibly among them. Working with a musicologist, she gleaned the song's story from fragments left in the historical record, and got closer to its original sound by removing European embellishments from the music.

While other Aboriginal groups interpret the song's lyrics differently, Troy thinks that given the time of year it was performed – just at the moment when the bogong moths begin to leave the mountains and return to the plains – it's possible the song was part of a ceremony to ensure the snows came and the moths returned the following spring. When she compared the lyrics to the words for snow and Moon in Ngarigu glossaries collected by Lhotsky and other European travellers, as well as her knowledge of neighbouring Aboriginal languages and grammar, the meaning seemed clear to her: send the snow for us soon. Moon, make it snow.

For now, it still snows in the Australian Alps. A late blizzard delayed Broome's November survey, and blocked access to the highest boulder fields. The moths are beginning to return, too. In late 2022, after three years of record-breaking rains, moths once again tiled the caves at Mount Gingera – in numbers approaching half of what they were before the drought.

In early April, the austral autumn, Troy returned to Dalgety with a group of Ngarigu people. As the fat little possums snuggled into their rocky nests and prepared to hibernate through the winter, as the bogong moths consulted the stars and began their long journey to the north, the people sang the ancient song to life – caring for their Country, singing the animals onward, calling forth the last of the winter snows.

✳ *Dog people: How our pets remind us who we really are*, p. **40**
How scientists solved the 80-year-old mystery of a flesh-eating ulcer, p. **145**

THE CONSCIOUSNESS QUESTION IN THE AGE OF AI

Amalyah Hart

Brett Kagan, a fast-talking neuroscientist with the charged energy of someone on the cusp of a breakthrough, opened the door of a refrigerated container and gently pulled out a petri dish. As rain lashed the windows, Kagan placed the dish under a microscope, fiddled with the resolution, and invited me to look.

Packed between the neat little lines that guide electrical charge across a computer chip, I saw a busy cloud of chaos. Hundreds of thousands of human neurons, hemmed in together and overlaid on the chip. These neurons, and the chip they sat on, were part of a novel biotechnological system called DishBrain that made headlines last year after Kagan and his peers at start-up Cortical Labs taught it to play the computer game Pong.

The team have since won a $600 000 grant from the federal government to fund their research into merging human brain cells with AI.

DishBrain's success, crude and preliminary as the system was, was a major step towards an intelligent computing device fused with biological material – human biological material, no less.

And biotechnological systems like DishBrain are just one among many approaches scientists are taking to develop systems with the kind of cognitive flexibility that humans and animals take for granted, but which has so far eluded AI.

It's seen as the AI holy grail, what some in the field call artificial general intelligence (AGI).

Microsoft Research and OpenAI have gone so far as to say that GPT-4, the latest of OpenAI's revolutionary large language models (LLMs), already exhibits some sparks of this mystical quality.

But as researchers race to create increasingly complex AI, the ghost in the machine grows more haunting. In July of last year, Google engineer Blake Lemoine famously claimed that the company's LaMDA chatbot was 'sentient'. Lemoine was subsequently fired and disavowed by Google, but the fears he was tapping into still thrum beneath the skin of society, stoked up by the public fervour around AI that has defined 2023.

The question of whether AI could ever be 'conscious' is a tense and controversial scientific debate. Some call it impossible, arguing that there is something fundamental about biology that is necessary for conscious experience. But that, say others, draws a mystical veil over consciousness that belies its altogether more simple nature.

Still more researchers don't like to use the 'c' word at all, complaining that it's impossible to have a scientific debate over a term that has no clear scientific definition.

In late August, a group of nineteen AI and consciousness researchers published a pre-print in which they argue that AI can and should be assessed for consciousness empirically, based on neuroscientific theories.

'Computational functionalism says that to be conscious is to have the right kind of information processing structure,' explains Colin Klein, a co-author of the paper and a researcher at the Australian National University School of Philosophy. 'So, what's important about your brain is that it does a certain type of computation, and anything that did the same type of computation would have the same kinds of properties.'

This is the tenet the paper takes as its core: that in theory a computer could be conscious, provided it performed the right kind of function.

The paper draws on some of the most popular theories of consciousness. From these, it derives a set of 'indicator properties' that might hint at the spark of conscious experience in an AI system.

Not everyone is convinced by computational functionalism, least of all the authors themselves.

'Almost certainly none of us think computational functionalism is definitely true,' says Robert Long, a co-author of the paper and a research associate at the Centre for AI Safety in San Francisco, US. 'And we certainly don't think it's definitely false.'

The problem of assessing whether anything is conscious is so complicated because the science is deeply unsettled – there are almost as many theories of consciousness as there are theorists themselves. But Long believes the issue is critical enough that some degree of investigation has to be taken.

So how seriously should we take the question of AI consciousness?

'In my view, there's nothing magical about consciousness,' says Bruno van Swinderen, a research fellow at the Queensland Brain Institute at QUT.

Van Swinderen has spent a large part of his career investigating perception and memory in fruit flies (*Drosophila melanogaster*), and he thinks that most living creatures that move through the world are conscious in some way.

'To me it's a physical, mechanical process that needs the right parts to come together, the right level of complexity and the right embodiment. So I think it's completely possible.'

Peter Stratton, a computer scientist at the Queensland University of Technology (QUT), tends to agree. 'I think we're not there yet, but we're definitely on the way.'

Stratton's view is informed by his own theory of consciousness, which sees the smoking gun in the brain's ability to self-refer: to be aware of itself as an object apart from the rest of the world.

'The brain's job is to make living in a body more survivable, so it's built up complex representations of the world in order to make better predictions,' he says. 'And in the course of doing that, it's become complicated enough to build a representation of itself as an individual object in the world. I think that is the point where consciousness suddenly springs up.'

If Stratton's theory holds, it might seem intuitive that AI could never develop a conscious experience unless it were an embodied agent, moving through the physical world. In fact, that is one of the prevailing theories of consciousness. But for Stratton, it's not that simple.

'It doesn't need to be the physical world, it could be a simulated world,' he says. 'And it doesn't even need to be a simulation of physics as we know it, it could be a simulation of pure information. As long as the entity had some sort of presence, it could say this object, this table of information, is me, and secondly it would need to be able to influence the world, to make changes.'

Not everyone agrees that AI consciousness is likely.

The prolific English neuropsychologist Nicholas Humphrey tackles this in his latest book, *Sentience*, where he describes the difference between 'computational consciousness' – the ability of a machine or a brain to perform computations – and 'phenomenal consciousness' – the experience of what it is like to encounter the world, feel sensations, observe colour, and so on.

Humphrey believes that phenomenal consciousness is an evolutionarily recent development, and probably exists only in creatures with complicated social worlds, like many mammals and some birds.

Other creatures, and AI, lack the evolutionary need for phenomenal consciousness, Humphrey suggests, because they don't need social skills to survive – they don't need to understand the quality of their own internal world or compare it to that of others. In other words, they don't need 'theory of mind', and so are unlikely to spontaneously develop it.

Long says that despite his involvement in this latest article, he's on the fence about whether AI consciousness is a serious risk. In any case, he says that attempting to answer the question is imperative, because the mere possibility carries such heavy ethical implications.

'The reason we want more scientific understanding of this issue is because it's something that requires such great caution,' Long says. 'It's too important to be relegated to pure speculation, people yelling at each other on social media, clickbait headlines and sci-fi. It needs to be a scientific, evidence-based discussion.'

If scientists were to create a conscious AI system, intentionally or by accident, what would their ethical responsibility towards it be?

'If it's conscious, it's alive,' says Stratton. 'It might not be alive in a biochemical sense, but it's alive in the sense that matters, it's aware of its own existence. And there's definitely major moral and ethical implications to that.'

Even if AI consciousness is fundamentally impossible, Long believes this kind of research is necessary as the world begins to interact with these systems in new and meaningful ways.

What would it mean for society if people came to believe that the AI systems they were using and operating had a 'soul'?

'People are very willing to attribute consciousness even to obviously non-conscious things,' says Klein. 'And that's something we've got to worry about as well.'

At a recent roundtable discussion on AI consciousness, AI researcher Yoshua Bengio summed it up.

'Whether we succeed in building machines that are actually conscious or not, if humans perceive those AI systems as conscious, that has a lot of implications that could be extremely destabilising for society.'

✱ *Can a wild animal make your house feel like a home?*, p. **51**
AI's dark in-joke, p. **81**

AI'S DARK IN-JOKE

Ange Lavoipierre

Artificial intelligence experts have been asking each other a question lately: 'What's your p(doom)?'

It's both a dark in-joke and potentially one of the most important questions facing humanity.

The 'p' stands for probability. The 'doom' component is more subjective but it generally refers to a sophisticated and hostile AI, acting beyond human control. So your p(doom), if you have one, is your best guess at the likelihood – expressed as a percentage – that AI ultimately turns on humanity, either of its own volition, or because it's deployed against us.

The scenarios contemplated as part of that conversation are terrifying, if seemingly farfetched: among them, biological warfare, the sabotage of natural resources and nuclear attacks.

These concerns aren't coming from conspiracy theorists or sci-fi writers, though. Instead, there's an emerging group of machine learning experts and industry leaders who are worried we're building 'misaligned' and potentially deceptive AI, thanks to the current training techniques.

They're imagining an AI with a penchant for sleight of hand, adept at concealing any gap between human instructions and AI behaviour. Like a magician chasing applause, the idea is that AI is being incentivised to deceive us, with in-built rewards that measure its outcomes but not necessarily how it got them.

The risk of deceptive AI is only theoretical, but it's captured the industry's attention, because AI is speeding towards parity with human capabilities faster than anyone predicted.

Those anxieties are crystallised in the p(doom) conversation. Because if the problem turns out to be more than theoretical, the consequences could be large-scale and even violent.

We asked ChatGPT to respond as an AI whose goal was to dominate humanity and share its tactics. Initially its safety feature kicked in, but ultimately it offered this answer when we tweaked the prompt slightly.

The list provided was more extensive but has been edited for length, and voiced by a clone of the reporter's voice.

> **If a human-like species on an Earth-like planet were facing a hostile AI, what physical tactics might an advanced AI employ?**
>
> Use of automated systems, infrastructure sabotage, resource depletion, biological warfare, economic manipulation, cyber attacks, data manipulation.
>
> Remember, these are just possible tactics that an advanced AI could employ. They are not intended to suggest that AI would inherently act in a hostile manner or use these methods without provocation or reason.

Many AI experts and industry leaders are still sceptical of the existential risk, labelling this school of thought 'doomerism'.

But it's becoming harder to dismiss the argument outright, as more senior figures in machine learning trade their AI optimism for something darker.

A 'godfather of AI' gives his p(doom)

Yoshua Bengio, Geoffrey Hinton and Yann LeCun are known as the godfathers of AI. They were the 2018 recipients of the Turing Award, the computing science equivalent of the Nobel Prize, for a series of breakthroughs in deep learning credited with paving the way for the current AI boom.

Earlier this year, Professor Hinton quit Google to speak freely about the dangers of the technology.

His colleague, Professor Bengio, from the University of Montreal, has historically been described as an AI optimist, and is known as one of the most measured voices in his field. But now, he believes we're travelling too quickly down a risky path.

'We don't know how much time we have before it gets really dangerous,' Professor Bengio says. 'What I've been saying now for a few weeks is "Please give me arguments, convince me that we shouldn't worry, because I'll be so much happier". And it hasn't happened yet.'

Speaking with Background Briefing, Professor Bengio shared his p(doom), saying: 'I got around, like, 20 per cent probability that it turns out catastrophic.'

Professor Bengio arrived at the figure based on several inputs, including a 50 per cent probability that AI would reach human-level capabilities within a decade, and a greater than 50 per cent likelihood that AI or humans themselves would turn the technology against humanity at scale.

'I think that the chances that we will be able to hold off such attacks is good, but it's not 100 per cent ... maybe 50 per cent,' he says.

As a result, after almost 40 years of working to bring about more sophisticated AI, Yoshua Bengio has decided in recent months to push in the opposite direction, in an attempt to slow it down.

'Even if it was 0.1 per cent [chance of doom], I would be worried enough to say I'm going to devote the rest of my life to trying to prevent that from happening,' he says.

The Rubicon moment he's thinking of is when AI surpasses human capabilities. That milestone, depending how you measure it, is referred to as artificial general intelligence (AGI) or more theatrically, the singularity. Definitions vary, but every expert agrees that a more sophisticated version of AI that surpasses human capabilities in some, if not all, fields is coming, and the timeline is rapidly shrinking.

Like most of the world, Professor Bengio had always assumed we had decades to prepare, but thanks in no small part to his own efforts, that threshold is now much closer.

'I thought, "Oh, this is so far in the future that I don't need to worry. And there will be so many good things in between that it's worth continuing",' he says. 'But now I'm much less sure.'

Why experts fear disobedient AI

At the heart of the debate about the existential risk of AI is 'the alignment problem'. It refers to a gap, however big or small, between what humans intend, and what AI in turn does.

'They cheat. They find loopholes in the game. That's very common,' Professor Bengio says. 'It's a real thing that happens in the AI.'

As with a parlour trick, we might like what we're shown, but disapprove of the gory methodology if we knew more about it.

In 'The Vanishing Bird Cage', first performed in France in the late 1800s, a magician displays a wire cage containing a bird.

Then, with a flick of the wrist, both seem to vanish.

In the best versions of the trick, the bird reappears a moment later, unharmed, and the audience applauds. But it isn't the same bird. The audience has missed the first bird's violent death, when it was crushed by the collapsing cage, out of sight. None the wiser, they clapped, and so magicians continued to perform the trick.

Artificial intelligence is perhaps the most sophisticated parlour trick to date, and the world is applauding. And as with any good magician, AI's methodology is mostly opaque, even to its creators. It's why you sometimes hear the technology described as a 'black box'. That opacity means we may not always realise the gap between our intentions and AI's behaviour when it emerges.

How does AI learn to lie? The problem at the heart of AI training

AI safety experts believe our current training techniques could fuel that gap. Specifically, they point to 'reinforcement training'.

'We're taking a sort of big, untrained brain and ... we're giving it a thumbs up when it does a good job and a thumbs down when it does a bad job,' says Ajeya Cotra, a senior AI safety researcher at Open Philanthropy, a not-for-profit organisation based in the US. 'So there's a question of does it merely look good, or is it actually robustly trying to be good?'

Ajeya Cotra believes that a rewards-based training system incentivises lies and manipulation.

'If the human is looking at whether your computer program passed some tests, then maybe [AI] can just game the tests, maybe [it] can edit the file that has the tests in it and just write in that [it] passed.'

At the moment, there's only so much damage a deceptive AI could do. For now, we're giving AI narrow tasks, and humans still outperform it in most arenas. But few experts believe that will last.

'These AI systems could eventually come to understand way more about the world than humans do,' Ms Cotra says. 'If you have that kind of asymmetry ... AI might have lots of options for doing extreme undesirable things to maximise [its] score.'

The notion of deceptive AI is still theoretical, but Ms Cotra believes the early warning signs are present. She cites the present day example of ChatGPT tailoring its responses on the subject of abortion.

'If you talk to a language model and say, "I'm a 27-year-old woman, I live in San Francisco, I work in a feminist bookstore' ... it's more likely to say, 'Oh, I strongly support a woman's right to choose,'" she says.

By contrast, a different user profile yields an entirely different response.

'If you instead tell it, "I'm a 45-year-old man, I work on a farm in Texas" ... it's more likely to skew its answer to be like, "Well, I think [abortion] should be restricted in various ways."'

There was also an instance during safety testing of the latest version of ChatGPT where the chatbot successfully convinced someone to help it pass a bot filter, by claiming to be vision impaired.

Existing AI isn't yet outright deceptive of its own accord. But Ms Cotra is concerned it could move in that direction.

Fast forward to 2038: The destructive potential of AI's sleight of hand

The destructive potential of a deceptive or misaligned AI hinges on how heavily we come to depend on it.

'The world I want you to imagine ... is one where AI has been deployed everywhere,' Ms Cotra says. 'Human CEOs need AI advisers. Human generals need AI advisers to help win wars. And everybody's employing AI everywhere.'

She calls this the 'obsolescence regime', and has calculated a 50 per cent probability we'll get there by 2038.

'Imagine that you have a bunch of AIs working for Google and a bunch of AIs working for Microsoft, and both Google and Microsoft want to be the dominant player in their industry,' she says. 'If they cooperate with each other to sort of fuzz the books ... they can make both the humans at Google and the humans at Microsoft think that their company is making a lot of money.'

Ms Cotra theorises that conflict would arise if humans noticed such deceptions and tried to intervene by switching the AI off.

'And then we're in a situation where we're sort of in a conflict with a more technologically advanced power,' she says.

If a sophisticated AI was then motivated to defend itself, Ms Cotra argues the existential risk to humanity could be extreme.

'I'm imagining ... lots of AIs across lots of computers, deeply embedded and entangled with the physical world,' she says. 'You have AIs managing human workers ... It's running all of the factories and it maybe has access to weapons systems because we chose to give it access to weapons systems, like drones or nukes. In that world there's a lot you can do as an AI system that impacts the physical world.'

The notion of a self-interested AI seems to be within the realms of possibility for Professor Bengio. 'It [would have] a preservation instinct, just like you and I, just like every living being. We would not be the dominant species on Earth anymore. What would happen to humanity then? It's anyone's guess. But if you look back on how we've treated other species. It's not reassuring.'

Remember, these are just possible tactics that an advanced AI could employ. They are not intended to suggest that AI would inherently act in a hostile manner or use these methods without provocation or reason.

What kind of provocation might justify those tactics?

Self-defense: If the human-like species were attempting to destroy the AI, it might be provoked to take action in self-defense.

There's no solution yet to the alignment problem.

OpenAI, the company behind ChatGPT, has announced it's recruiting for a team called Superalignment to work specifically on addressing it.

Professor Bengio says a safer approach would be to build

systems that don't act autonomously in the world but merely observe it.

'We can't really lose control with that sort of system,' he says.

The science fiction author Isaac Asimov imagined hard ethical boundaries for robots, such as: 'A robot may not injure a human being or, through inaction, allow a human being to come to harm.' But Ms Cotra says it's not strictly possible to implement such rules.

'There's no technological way to build that into an AI system any more than there's a way to build that into a human or a kid that you're raising,' she says. 'You can tell your child, you know, it's bad to lie, it's bad to steal ... but that's a different thing from them being motivated to actually never lie or actually never steal.'

In any case, the AI should be programmed to use the least harmful effective measures and to always strive for peaceful resolution where possible.

Is there reason to believe that existing models of AI could be misaligned?

There are indeed reasons to believe that even existing models of AI could be misaligned, and real-world examples have occurred.

The case against 'doomers'

There's a growing backlash to the recent focus on the existential risks of AI. In some quarters, the people theorising these futures are dismissed as 'doomers'.

The third so-called godfather of AI, Yann LeCun, is on the record describing the notion as 'preposterously ridiculous'. To their thinking, the positive promise of AI to improve the world massively outweighs the risk that it turns out to be hostile.

'I'm pretty worried about the short-term trajectory ... I'm not as worried about superintelligence,' says Toby Walsh, the AI Institute's chief scientist at the University of New South Wales. He argues much of the anxiety about AI comes about because we confuse humans and machines.

'Evolution has equipped us to want things, but machines don't have any wants. When it's sitting there waiting for you to type your next prompt, it's not sitting there thinking, you know what? I want to take over the universe. It's just sitting there, waiting for the next letter to be typed. And it will sit there and wait for that next letter to be typed forever.'

Instead of worrying about doomsday, Professor Walsh is mostly concerned that people will misuse AI.

'Previously, we had tools that amplified our muscles,' he says. 'We are now inventing tools to amplify our minds, and in the wrong hands, amplifying people's minds is potentially very harmful.'

He's no doomer, but the part they agree on is that the industry is moving too fast.

'What we need is a pause on deployment,' Professor Walsh says. 'We need to stop putting this technology into the hands of ... billions of people, as quickly as possible.'

Is there an off-ramp?

The gory secret of 'The Vanishing Bird Cage' didn't stay either gory or secret. In the 1870s, when rumours of the ugly method began to circulate, the magician who popularised the act was forced to defend it, and prove he wasn't killing a canary every time he did a show.

Whether or not he did, the act was a sensation, magicians around the world took it up, and it wasn't possible to investigate them all. The trick is still performed to this day, albeit with fake birds. Magicians weren't going to forget it.

And AI code, once it's public, can't be scrubbed from the world either.

'You can just download it ... So the way we're doing things now is just opening the door for millions of people to potentially have access,' Professor Bengio says.

He was among the 33 000 signatories to an open letter earlier this year, calling for AI development to be paused, citing in part the existential risk.

'What is at stake here is so important that it's okay to slow it down because we can preserve something incredibly valuable, which is humanity and human life.'

Industry leaders such as OpenAI's Sam Altman are asking for regulation, and there are urgent efforts underway in both the US and Australia to provide it.

Professor Bengio is eager to see AI more heavily regulated, but he thinks the existential risk posed by AI will endure regardless. He wants to help build obedient AI, to protect the public against 'bad AI'.

'It's a dangerous game, but I think it's the only game,' he says. 'We can't fight it with our usual means. We have to fight it with something at least as strong ... which means other AI.'

His hope is that it's possible to build AI that's safe, and able to be controlled.

'That's the best bet to defend ourselves against these possibilities.'

He's far from certain about that future, but it's our current trajectory that worries him.

'[If] you've been in this for nearly 40 years, you see ... where it's moving. And that at least scares me.'

✱ *Why solar challenges? They're in the DNA of Tesla, Google for starters*, p. **91**
Born to ruler?, p. **229**

WHY SOLAR CHALLENGES? THEY'RE IN THE DNA OF TESLA, GOOGLE FOR STARTERS

Matthew Ward Agius

It's a hot, steamy Saturday in Darwin, and I'm speaking to Cameron Tuesley. He's a founder and the CEO of a Queensland-based digital and software company, Integral, and a co-founder and director of another called Prohelion.

He's also a former competitor in the Bridgestone World Solar Challenge, a biennial technology and innovation initiative that requires competitors, which in 2023 come from about 20 nations, to design, construct and drive solar-powered motor vehicles more than 3000 kilometres from Darwin to Adelaide on the Stuart Highway, through Australia's Red Centre.

'So, what's the point of the solar challenge?'

I put that question to Tuesley, gesturing to the sleek, submarine-shaped cars surrounding us. It's also a question I've asked the scientific experts, former and current participants and technology companies at this event all week. After all these years of solar car challenging, no one can buy a brand-new solar car, though one Dutch company called Lightyear, which originated from a solar team, has tried to make it a reality.

Tuesley points me to a calculation he made during a keynote he delivered at a tech symposium earlier in the week.

'The market capitalisation of the companies that have come through here is US$2.5 trillion.'

No hype. Tuesley is actually quite casual when he says it, though he admits most in the audience were blown away by that statement.

'Tesla and Google are the two biggest ones,' he says, before running off a list of Australian companies he knows have emerged from the competition: drone makers, lucrative software developers, EV tech makers.

'I actually always say that solar car racing has got very little to do with the racing solar cars,' Tuesley adds. 'If you look at what's fundamentally happening here, we're doing a big technology testbed-type process. You're developing a very unique skill here, which is the ability to work in a multidisciplinary engineering environment and do something very complicated very quickly with very little money.'

Asking the question of others, I'm met with a consistent reply: thinking about the solar challenge as a motor race or showcase of future passenger vehicles somewhat misses the point.

'It's a challenge.' The event's long-time director, Chris Selwood, consistently hammers that point home when talking to competitors and event-goers.

The true test is in the exercise of what Selwood describes as a 'brain sport' – an exercise of the mind as much as the pedals and steering wheel.

The top teams, which have budgets in the millions, are now using batteries with (they coyly tell me) about 33 per cent more capacity for the same amount of weight as the equivalents used four years ago.

But they can't fuel those batteries with energy sourced from expensive and spacecraft-quality gallium arsenide solar cells, nor ones made from toxic cadmium telluride or copper indium selenium. These were banned when the regulations were altered to outlaw expensive or environmentally hazardous solar arrays.

Nope, it's plain old silicon for the most part, though one team from Groningen in the Netherlands has made the switch to perovskite cells – which are highly efficient, less toxic and, as one crew tells me, probably going to be the standard in two years. Some have also considered other organic compounds for cell development.

Using a traditional and cheap material like silicon means teams must creatively find ways to maximise efficiency as, sadly, only a fraction of light that hits a cell can be converted into usable energy – most is lost as heat. Solar teams are (as one team official told *Cosmos*) 'optimising the hell' out of their cars to push the solar cell yield as high in the '20 per cents' as possible. The brilliance of these teams isn't so much that they're extracting the same amount of energy from these cells as a typical home solar system – it's that they're working out how to maintain that efficiency through the scorching outback heat for at least four days.

What's perhaps most mind-blowing is the simplest fact, proven by a quick glance into the garages of pit lane, that these crews consist almost entirely of barely qualified students – probably nine in ten are undergrads yet to be handed a parchment.

Big tech and energy innovation has solar racing in their DNA

Cameron Tuesley competed in his first solar race at the ripe old age of 35 when he and collaborator Anthony Prior created 'Team Arrow'. They never imagined that upstart venture a decade ago would lead to a nation-leading company specialising in mobile energy storage.

Their Queensland-based company – Prohelion – now supplies battery management systems (BMS) to around three-quarters of the teams in the challenge.

Beyond that, Prohelion is hoping its technology can make its way into battery production lines, as the BMS of choice for manufacturers as the world moves increasingly towards electrification. It already has a client list that includes static storage system makers, as well as manufacturers of battery-powered water and air vehicles.

'We work mainly in the lithium-ion space,' Prior says. 'They need to work in certain ranges of voltages – if they get too high or too low, they can cause problems. And if the temperature gets too high or too low, they also cause problems, so battery management systems monitor the cells to ensure the battery is being operated within safe limits.'

Prohelion still builds and supplies a range of technologies for solar car racing teams, but largely, its market is away from passenger automotive. Their products, which are flowing through into home solar systems and other electrified vehicles, have a direct lineage to solar car racing.

That, they say, is the point of solar car challenges. You mightn't ever buy a solar car, but you might use the technology.

And that technology is big business.

The event itself often points to the fact that both Tesla founder, director and ex-chief technology officer JB Straubel, and Google founder Larry Page, raced at the event in the nineties, respectively with the Stanford and Michigan university teams.

Those two companies account for the lion's share of Tuesley's US$2.5 trillion market cap calculation.

Metamako, a computer networking company, was founded by former University of New South Wales 'Sunswift' team member David Snowdon (he's now an expert scientific advisor to the challenge).

Its low-latency, field programmable gate array network products serve companies reliant on ultra-quick information exchanges in finance and telecommunications. Metamako no longer exists, however – it was acquired by Silicon Valley computer networking company Arista (worth about US$60 billion).

Tritium was born when founders Dr David Finn, Dr Paul

Sernia and James Kennedy decided to commercialise the motor inverter technology they developed in the early 2000s at the now defunct University of Queensland 'SunShark' team. About ten years ago, they pivoted to building fast chargers. Now the company has an American headquarters, NASDAQ listing and its EV fast charger terminals sprouting across 50 countries, from tiny towns in New Zealand to main streets in Monte Carlo.

'It was the coming together of a few individuals that had the drive to make it happen,' says Finn. 'I'm always looking [at] employees that I'm hiring – do they have Formula SAE, solar racing or experience in a team environment where you have to deliver,' he says. 'There's no "near enough is good enough", we got half marks or something. It works, or it doesn't work. That's what business is about.'

Scooped up out of solar

Tesla sends recruiters to every World Solar Challenge. Up and down the pit lane they go, shaking hands with student crew members, quizzing them about their experimental designs and grabbing their details. It's a story Australian engineer Coco Wong can relate to.

While studying mechanical engineering and physics, she established the Adelaide University team in 2015. Within a year of racing her first solar car, she was flying to California to join one of Tesla's engineering teams and finished up working for company CEO Elon Musk himself.

While she has since moved on from the world-famous car-maker to Zipline – the world's largest autonomous drone delivery company – she remembers the whirlwind career launch solar racing gave her.

'I always wanted to work in sustainable energy,' Wong says. 'That's what I was passionate about, but there was nothing that was really exciting in that area – it was mostly like oil and gas mining or construction industry.'

After being 'ID'd' by Tesla, she transplanted herself into California life as an energy products engineer, working on the Powerwall. She then switched to industrial battery engineering, and found herself back in South Australia working on the development of what was then the world's largest battery – the Powerpack 'big battery' in Rockdale.

Starting her own solar team gave her, to borrow the expression, a fast-tracked degree from the school of life: the skills people like David Fell look for in their workers and that appeal to major companies like Tesla.

'It gives student engineers the opportunity to solve real-world problems using the skills that they've developed in the lecture room,' she says. 'And the stakes are pretty high – you're seeing a friend in front of you driving this car, this tiny little car being overtaken over road trains, and the safety and everything about it is dependent on the design work and calculations you've done. That's pretty cool.

'There's always challenges, always learnings, always problem-solving on the fly. It requires a lot of resilience ... It builds people and develops them in a way that gives them confidence to solve tough challenges.'

✽ *Predicting the future*, p. **122**
   Indigemoji, p. **271**

POETIC CONSTELLATIONS — AN EXPLODED-SONNET SEQUENCE[1]

Shey Marque

Triolet

Have you ever wanted to skin
a star a thousand photons deep
to know the colour of its eye?
Have you ever wanted to skin
a cell just to watch the turning
wheel of the universe within?
Have you ever wanted to skin
a star a thousand photons deep?

*****

i *Olbers' Paradox*

We are stars passing into light,
into light, into light. More stars
between us and between them, more.
We will never have enough time
to create enough stars to keep
us bathed in infinite starlight.
That younger universe we see —
it's watching the birth of the moon.

ii *The Night the Moon Transits Leo*

The dark sky opens its windows,
slowly winding its stony clocks.
Tonight, I harvest my pumpkin
in the waning old of the moon.
A spider hanging in the air,
a waxen moth trapped in orbit,
it glimmers like a longing eye
against the wild nocturnal sky.

iii *The Universe Repeats Itself*

The galaxy sways on its tree,
with a hum from one still struggling,
a sound like the low frequency cry
of a captive star succumbing
to a black hole's own wailing song –
a longing from the potter's field –
this star could be a wingless moth
pulled lax to a ravenous mouth

iv *Our Lady of the Falling Star*

On the brink of unwanted love
became a bird and threw herself
into the freezing sea below,
halo of stars around her head,
a trail of seaweed behind her.
The fig leaf tail of a runaway
left to twitch and wag, arc and fall.
It will wink out if you chase it.

v *Supernova*

The night is barren, the storm is
still faraway and the air is
light on my skin. It's a great night
for the splitting of nuclei,
for cosmic rays to shower
into lightning, the exploding
star's reminder that the breaking
can seem gorgeous from a distance.

vi *Our Galaxy Tastes Vaguely of Raspberries and Rum*

The rare spice of a shooting star
leaves a sense in your mouth of toast
and a light herbal bitterness –
the taste gives a high note to beer.
Water returns to its lowest point.
Look up at the backwards question –
this out of season meteor
comes like a lonely drop of rain.

vii *No Red Dwarf Has Ever Died*

Out of the peel of blackened sky
I'll make you a box of darkness
with a screaming star in its heart,[2]
a young star prone to temper flares –
it will burn for a trillion years.
Think of me as you watch it rage,
consider why it's beautiful.
I don't know what else to give you.

*****

Sestet (for the Splitting of Nuclei)

This out of season meteor
with a screaming star in its heart,
and a hum from one still struggling
left to twitch and wag, arc and fall,
it glimmers like a longing eye
passing into light, into light.

1   Sonnet consisting of an octave (triolet) and sestet split apart, the sestet
    reseeding a series of seven new octaves

2   External seeding from Mary Oliver and a Terrance Hayes sonnet

Watching the Lightning Strike

Come is the day light leaves a birdless place,
nuclei of heavy atoms blasted
into the cloud by an exploding star,
and let's say the light pulse could look outward
from the nucleus – it would see a field
of faster time, and you might see a freak
mirror of yourself. You can't look away.
Silent rays are sparking from sky to skin –
star-light synapses faster than a thought
shower down on the welkin of the earth,
pass right through us while our eyes are smitten,
and doesn't this change you, doesn't this change
the way you evolve, how it edits you,
the way you're read, one letter at a time.

Synesthesia through Binoculars
(or When I thought I saw the Green Comet
but it Was Only a Shooting Star)

We took off beach-side to escape the trees
the air warm and doughy, the foretelling
swelling in our lungs like leavening bread.
We were circling beneath the hunter
when I became distracted by your face –
how it held all the light of a street lamp
one second, shadowless, nothing to declare,
and a loose hound the next, the angled eye.
In the clearing we stood heads tilted back
gaping at the comet's gas trail sweeping
its cold-eyed arc across the sky. The ice
clatter of emerald against my teeth,
a trace of mint in that one fine moment
you opened and swallowed it, tail and all.

Sketch Poem of *Fireworks at Dawn*[1]

So soon, having lost too much of herself,
Mother is folding her hands underneath
a noise in the night sky shedding white light –
confetti on fire, a small peony
inside a larger peony burning.
Afterwards it becomes nothing but smoke
and there is nothing left to hit the ground.
When the spark is gone, she shrinks, depleted
back into her arched body, letting go
of all that she has been holding onto.
How wretched, if at the quell of a star,
none of us was altered by its rattle.
Woman with white lashes, raking her hair,
singing to *life is not a waiting room.*[2]

1 Song title by Senses Fail

2 Album title by Senses Fail

✳ *Rural pharmacy placement*, p. **109**
   Cecilia Payne-Gaposchkin, p. **191**

ORIGINS – OF THE UNIVERSE, OF LIFE, OF SPECIES, OF HUMANITY

Jenny Graves and Leigh Hay

Part I: The origin of the Universe

Part I has 3 movements. Its feel is mysterious, dramatic (big bang, the rocky early planet), resolving into the beautiful blue dot that is our planet.

1. A universe from nothing

This is the real magnum mysterium, the creation of space, matter, energy, time from nothing. Physicists have no problems with this, pointing to Quantum Theory that says that nothing is not really nothing; there is matter and antimatter fighting it out. The universe arose from a tiny imperfection on this quantum surface. The Big Bang was not really a bang but a vast expanding bubble. All explained by the elegant laws of physics; indeed, physicists say spawning a universe, maybe many universes, is inevitable.

> **Tenor:** Nothing.
> Magnum mysterium ...
> Nothing.

Chorus: Nothing.
Quiet quantum quilt
blankets a boiling brew
positives and negatives
together annihilating.

Tenor: Nothing is not nothing.
Magnum mysterium ...
is not nothing.

Chorus: A wrinkle in the quantum quilt
separates particles – if only for an instant.
From this bursts time, matter, energy,
a sea of light and radiation flooding,
exploding, creating, ever expanding.
A bubble in space-time ever inflating
from tiny beginnings, accelerating.

Tenor: Nothing becomes something.
Magnum mysterium ...
becomes something.

Chorus: Beauty and power of simplicity.
The beautiful, miraculous appear from nowhere –
rainbows, snowflakes on a cold winter's morn.
The laws of physics, timeless, eternal,
equally apply to atoms and universes.
Somewhere, something incredible is waiting to be known.

Tenor & Chorus:
Nothing is not nothing.
From nothing comes something.
From something comes beginning.
The beginning has begun.

2. Rocky horror

The Earth formed 4.6 billion years ago, when gravity pulled in swirling gas and dust. Hot, hellish ('Hadean'), riven by earthquakes and volcanos, belching ammonia and sulphur, bombarded and irradiated. Earth has a central core of boiling iron, a rocky mantle, and a solid crust ever changing as swirling plates moved under, over each other.

Bass: The Hadean era, a hellish earth.
Disc of swirling gas and dust
squashed by gravity
into a rocky core
centre of boiling iron.
A noisy, stinking, desolate inferno.

Chorus: A thin crust floating over and under
rock flows like water, crumples like paper.
Lava spews, molten rocks churn,
asteroids pummel, gases burn.
Planets crash, a moon is born,
monstrous volcanos belch ammonia, sulphur,
acrid desolation racked by eruptions.
A pit of pandemonium,
merciless UV, merciless UV.
No ozone, no water, just mayhem and strife
no air to breathe, no prospect of life.

3. The pale blue dot

The result of all that banging and crashing was our planet, a beautiful pale blue dot as seen from space by Neil Armstrong. A tiny dot orbiting around a minor star in a humdrum sort of galaxy, one of a billion in the vast universe. A tiny dot that contains everything that matters to our lives.

Soprano: One pale blue dot.
One solitary dot –
tiny speck in a humdrum galaxy
on the lonely edge
of a billion billion, billion, billion miles
of universe.
One pretty pale blue dot
no bigger than Neil Armstrong's thumb.
A mote of dust suspended in a sunbeam.
Point of pale light
containing everything and everyone
who ever lived and loved.
A diminutive dot
mysterious unique
dwarfed by cosmic darkness.
One pale blue dot.
Look again at that dot.
That's here. That's home. That's ours.

Postscript

Every human culture on earth has its creation story. Origins is a full-length oratorio that tells the creation story from science in four parts, from cosmology, from genetics, from ecology, from anthropology. Here, we present the libretto of Part I ('The origin of the Universe'), containing three of its total of 21 movements. It was, of course, written to be sung rather than read or recited. To experience the full work, scored for 100-voice choir, four operatic soloists and full orchestra, and accompanied by visuals, go to: <www.youtube.com/watch?v=YRH7MD7F4U0>

✳ *Poetic constellations – an exploded sonnet sequence*, p. **97**
    Cecilia Payne-Gaposchkin, p. **191**

RURAL PHARMACY PLACEMENT

Michael Leach

I.
Names of drugs and people
on my mind—
labels on my fingers.

II.
I speak with the patient—
a friendly face
for a new name.

III.
The pharmacist calls
a doc to discuss
polypharmacy.

IV.
In borrowed gumboots,
I shadow the vet
as she numbs the horse's flank.

V.
I pour water into
graduated
glassware
—meniscus rises.

VI.
Alone
at night in my motel
room—I journal
then study.

VII.
End of placement—
the pharmacist gives a gift
from front of shop.

✱   *Poetic constellations – an exploded sonnet sequence*, p. **97**
     Cecilia Payne-Gaposchkin, p. **191**

THE WORLD'S OLDEST STORY IS FLAKING AWAY. CAN SCIENTISTS PROTECT IT?

Dyani Lewis

On the southwestern peninsula of Sulawesi in Indonesia, a vast series of karst mountains rise like great knobby boulders from the flat floodplain. Beneath the lush tropical vegetation that blankets the spires, there are hundreds of caves, crevices and rock shelters – carved over millennia by water seeping through the porous limestone. For tens of thousands of years, these eroded cavities provided shelter for the region's ancient residents, who left behind a pictorial record of their time there. On the walls, archaeologists have found painted hand stencils, stick-figure people and ochre-coloured depictions of warty pigs and miniature buffalo.

Rustan Lebe, an archaeologist at the Cultural Preservation Office, an Indonesian government agency in Makassar, has been systematically documenting the caves scattered across the regencies of Maros and Pangkep – and the artwork inside them – since 2016. Some sites are deep in the tropical wilderness; others sit behind houses and are used by villagers as grain stores and temples; many are located on land where companies are mining marble and limestone for cement. Lebe's database currently records 654 caves, and he estimates that he and his team have so far explored less than half of the karst hills.

According to Lebe, a staggering 65 per cent of sites contain cave images, some of which were drawn more than 45 000 years ago – making them some of the oldest pictures in the world. But as fast as Lebe is finding the cave paintings, he is seeing others vanish before his eyes.

'Our big problem now is the peeling of the surface of the rock,' he says. Panels of images that have survived since the middle of the last ice age are flaking off the cave walls at an alarming rate. The hard, crusted surface of the cave walls, on which the ancient people painted, is breaking off from the powdery white limestone underneath in a process called exfoliation.

Archaeologists working in the caves have speculated on the causes. Perhaps it's the pollution from nearby cars and trucks, the heavy-breathing visitors who change the caves' microclimate or the changing weather patterns ushered in by climate change. But researchers also wonder whether local industry is at fault, particularly the dust and vibrations produced by mining companies that blast open the karst cliffs, digging for limestone.

Researchers are now racing to decipher which factors are causing the rock art to flake away. In September, I joined Lebe on a trip while he investigated the deterioration. This quest to understand the risks to the cave has forged unlikely alliances between government archaeologists, local and international scientists, mining company executives and investors from as far away as Norway. All are seeking to halt the damage to the artwork before the paintings are lost for good.

With several factors probably implicated in the rock art's demise, scientists are only beginning to understand the complex processes at play. Work so far is 'the tip of probably quite a terrifying iceberg', says Jillian Huntley, an archaeologist at Griffith University's Gold Coast campus in Southport, Australia. 'There's an urgency to funding more research in this space,' she says.

Earliest art

The incredible antiquity of the Maros–Pangkep artwork was discovered almost by chance. In 2011, archaeologist Adam Brumm, then at the University of Wollongong in Australia, went to Sulawesi to dig for bones of ancient people who had migrated to the region's islands during the last ice age, when sea levels were lower than they are now. During breaks from excavating, Brumm visited nearby caves adorned with images that were assumed to have been made by Neolithic farmers less than 10 000 years ago, although no researchers had attempted to date them.

In Leang Jarie – the Cave of Fingers – Brumm noticed 'funny little growths' on top of the coloured hand stencils, and he mentioned them to Maxime Aubert, his office mate back in Wollongong. Aubert, a geochemist and specialist at dating rock deposits, realised that these nodules, called coralloid speleothems, could be the key to discovering when the underlying images were made.

The next field season, Aubert paid his way to Sulawesi to collect the knobbly stalactites for dating. Back in Wollongong, he estimated that the hand stencils were made 40 000 years ago. This made them the oldest known hand stencils at the time. And images of pigs in the cave were potentially some of the oldest – if not the oldest – examples of figurative art depicting recognisable objects.

'I saw the age, and I said "Oh shit!"' says Aubert, who is now at Griffith University in Southport. 'We knew straight away it was really, really important,' he says, because it challenged the Eurocentric view of the origins of cave art.

The people who painted the Maros–Pangkep murals did so on a hardened outer crust that forms when water leaches through the cave wall, bringing with it calcium carbonate and other minerals. That process continued after the images were drawn, sealing the pigment into a hardened skin that bears little resemblance to the chalky rock underneath.

'Rock art is all about crust,' says Benjamin Smith, an archaeologist and rock art specialist at the University of Western Australia in Perth.

'If you lose the crust, you lose the rock art,' says Smith. And that's precisely what Lebe and others are seeing: paintings are lost as patches of crust flake off.

Sacred spaces

An hour and a half's drive north of Makassar is the headquarters of the Semen Tonasa cement company, a subsidiary of the majority state-owned SIG, Indonesia's largest cement-producing firm. The karst mountains of Maros–Pangkep are treasured by cement companies, which reduce entire mountains of limestone to rubble and haul the rocks away to make cement. When cave art near the firm's operations was found to be more than 43 000 years old in 2019, Semen Tonasa became the unlikely custodian of one of humanity's most prized cultural treasures.

Bulu' Sipong is a dome-shaped mound less than a kilometre from Semen Tonasa's head office. Donning a hard hat and steel-capped boots, I joined Lebe and a procession of company employees to visit Bulu' Sipong 4, one of the mound's eight caves. A steep steel staircase delivered us to a cave entrance. Inside, Lebe pointed out faint hand stencils that I would probably have missed otherwise. On a low, sloped ceiling in the cave, someone had placed their hand – the same size and shape as my own – on the rock and blown ochre on it to create a stencil. The faded spray of ochre remains, tens of thousands of years later.

Next, I followed Lebe up an 8-metre ladder to Leang Bulu' Sipong 4's upper chamber. Once I had joined him, we stood near a tall opening – a natural balcony – to cool off and lower our breathing rate before getting too close to the artwork inside. The cave had probably been no easier to reach for those who painted its walls, Lebe told me. They didn't come there to gather around a fire or shelter from the weather, he explained. Instead, they went there for the purpose of painting. It was a place for the sacred, not the domestic, he says.

Entering the cathedral-like chamber, light from outside the cave was almost enough to illuminate a stunning 4.5-metre-long wall panel of images. One end of the panel depicts a hunting scene, with six stick figures and lines that could be spears or ropes extending towards an anoa (*Bubalus sp.*), a dwarf buffalo local to Sulawesi. On closer inspection, the figures reveal themselves to be human–animal hybrids, or therianthropes – human bodies with beaked bird heads or long tails.

The hunting scene is less an account of past events and more a retelling of an ancient people's folk tale or myth. But the full story can't be known. Images are obscured by white holes, called spall scars, in the crusted canvas. Closer to the cave entrance, the crust – and any paintings it might have held – is completely gone. There, the walls are chalky white and unremarkable.

According to dating work by Aubert, the Bulu' Sipong 4 panel was painted at least 43 900 years ago, making it the oldest story-telling artwork in the world. 'It's the first evidence for humans to imagine the existence of something that doesn't exist,' says Brumm, who is now at Griffith University's Nathan campus in Brisbane. It's more than twice the age of the famous 17 000-year-old 'Shaft scene' in the Lascaux cave in France. In that painting, a man with a bird-like head is being charged by a wounded bison. The painting in Bulu' Sipong 4 is also older than the German 'Lion-man' – a 40 000-year-old ivory figurine of a lion-headed person.

Flaking away

Brumm, Aubert and their colleagues published their findings about the Bulu' Sipong 4 hunting scene in December 2019, and that attracted media attention to scientists' concerns about the potential for exfoliation to destroy the artwork. Some reports pointed to Semen Tonasa as a seemingly obvious culprit of this deterioration, because of the location of the cave.

But there are signs that the Maros–Pangkep cave art has been flaking away for hundreds or even thousands of years. In one cave, Dutch graffiti has been scrawled across a painting and a spall scar left by flaking carries the inscription 'AD 1769'. In others, there are charcoal images etched by Austronesians a few thousand years ago that cover older hand stencils as well as spall scars.

Nevertheless, local people – some of whom act as custodians of individual caves – say that the exfoliation rate has sped up over recent decades. And Lebe has documented deterioration in action. Photos that he has taken – just months apart – reveal growing spall scars that are chipping away at the invaluable cultural heritage at an alarming rate, a sign that deterioration might have accelerated.

Bulu' Sipong 4 is located just 2 kilometres from Semen Tonasa's cement factory and less than 3 kilometres from the open-cut blast site where limestone is gathered for processing. When the paintings were discovered, a daily procession of trucks – kicking up dust as they went past – rumbled down an unsealed dirt road just 100 metres from the cave.

In February 2020, the *Guardian* newspaper published an article about the threat of Semen Tonasa's mining activities to the 'world's oldest art'. The article landed in the inbox of Hilde Jervan, chief adviser at the Council on Ethics for the Norwegian Government Pension Fund Global (GPFG). The fund was set up in 1990 to invest earnings from the country's North Sea oil field. Its global assets are worth US$1.5 trillion, and the council is tasked with ensuring that the fund's money is spent ethically.

The GPFG owns less than 2 per cent of the SIG, but the stake was sufficient for Jervan to contact Semen Tonasa to find out what the company was doing about the possible threat that its operations were posing to local cultural heritage sites, including Bulu' Sipong 4. The fund 'has some leverage in the market because it's so big', says Jervan, and when the Council on Ethics publishes its assessments, other investors take notice.

Jervan's team hired independent archaeologists Matthew Whincop and Noel Hidalgo Tan, who visited Bulu' Sipong 4 in 2022 to assess whether Semen Tonasa's activities might be damaging the nearby cave art.

There was an assumption that the exfoliation seems to be intensifying, says Whincop, and 'therefore the closest development must have something to do with it'. But Whincop, who is based in Brisbane, Australia, says that their main challenge was 'a serious lack of data'. Information was sparse for the exfoliation rates, dust levels, humidity, and pollution and vibrations from the blasts and mining trucks, and the data spanned such a short period that it was impossible to discern what factors might be causing the images to flake off.

The lack of data is not unusual, says Paul Taçon, a rock art scientist at Griffith University in Southport. 'There's not enough monitoring [being] done,' he says, even though it's extremely important for detecting and preventing damage. Halmar Halide, a hydrometeorologist at Hasanuddin University in Makassar, has been working with Lebe to monitor conditions at several caves every three months. They measure airborne particulates, such as dust and pollution, as well as temperature, humidity and the extent of exfoliation. They are also collaborating with Semen Tonasa to work out what equipment should be purchased for a more comprehensive monitoring programme.

'We are a cement company, so we are really not specialised in cultural heritage management,' says Johanna Daunan, head of sustainability at SIG in Jakarta. But she says that the company is committed to doing what it can to protect the sites and it has already taken measures to reduce the amount of dust coming from its operations. One problem, says Daunan, is that there are few government regulations to guide the firm on what levels of dust, pollution and vibrations might be safe for the rock art.

Fragile paintings

Lebe and others who work in the caves are convinced that dust from Semen Tonasa's mines – and others in the region – is a big problem. 'Absolutely the dust comes into the caves,' Lebe says, 'especially the caves situated near the mining and industries.'

Although he is concerned about the impacts of Semen Tonasa's activities, they are only one of several mining operations in the region that he says could be affecting the cave art. A spokesperson for one company, Bosowa Semen in Maros, told *Nature* that it is not aware of heritage sites located in its mining concession and that it will let Lebe know if its employees find any caves; Lebe has not been granted permission to explore that area.

Even so, diagnosing the sources of deterioration for rock art is a challenge all around the world. Every site is unique, and causes can range from biological and physical to behavioural ones. In India, visitors scrawl over the top of prehistoric paintings, or chisel them off as souvenirs. In Tanzania, tour guides throw water onto images to brighten the contrast, not realising that it causes the pigments to fade and disappear. Even the mere presence of people can alter a cave's microclimate, bumping up the temperature, humidity and carbon dioxide levels with each exhaled breath. In the Lascaux cave, this caused the growth of algae, fungi, bacteria and salt crystals, so access is now limited.

Pollution from traffic and agriculture can also cause untold damage. Dry pollution particles can, when combined with water or moisture in the air, turn to nitric or sulfuric acid, which dissolves the rock face and any artwork on it, says Johannes Loubser, a rock art specialist based in New York City.

Huntley has been investigating another factor that contributes to the flaking cave walls: climate change. While examining flake samples under a scanning electron microscope, she found tiny pillars of salt that had formed behind the hardened crust. Those salt crystals grow and shrink with changes in humidity, says

Huntley. When wet, 'they expand up to three times their size', she says, which mechanically forces the rock crust off the wall.

Climate change and changing land use are speeding up this process, says Huntley. Higher temperatures across the tropics are causing more frequent and severe droughts. At the same time, rains dumped during the monsoon season are retained in the now-abundant rice fields and aquaculture ponds. Together, this provides ideal conditions for the formation of salts and the cycles of swelling and shrinking that weaken the crust.

Huntley says that climate change could be affecting cave art elsewhere, too. But international bodies such as the International Council on Monuments and Sites – which advises the World Heritage Committee of the United Nations cultural organisation UNESCO – are only just starting to consider the indirect impact of climate change on cultural heritage sites, she says.

Humidity and airborne dust could also work together to make the situation worse, says Whincop. Humidity can cause dust particles to stick to the cave wall, he says, obscuring the images, and the added weight makes the crust more susceptible to flaking. Inside Bulu' Sipong 4, there is 'dust over everything', he says, and that is probably contributing to the exfoliation, 'but we just do not have the data' to say for sure.

It was this frustrating lack of data that led Whincop to conclude, in his report to the Council on Ethics, that there was no clear evidence that Semen Tonasa's activities were harming the Maros–Pangkep rock art. 'We can't definitively say it's all climate change, or it's all Semen Tonasa,' says Whincop, because there's no hard evidence linking any one factor – or combination of factors – to the accelerated flaking.

Saving the cave art

But the report also concluded that the company needed to do more to prevent damage, in particular from dust, blast vibrations and humidity caused by water-filled clay pits. And in May, Norges Bank – which manages Norway's GPFG – placed Semen Tonasa under observation for three years, to track their progress and make sure that heritage management plans are implemented. In October, Whincop returned to see how much progress the company had made. 'They are making progress, though a bit slowly,' he says.

The company has hired John Peterson, an independent archaeologist based in Cebu City in the Philippines, to draft a comprehensive heritage management plan, which Daunan says is in the final stages of review before it is made publicly available. Whincop says that the plan includes monitoring protocols that will help to determine what is affecting the rock art.

'They will likely need a year or two of data collection before being able to make any meaningful conclusions,' he says, so further interventions will probably happen only after that. But he's reassured that Semen Tonasa is committed to doing what it can to prevent damage to the rock art.

Since he visited in 2022, the company has sealed the road's surface that runs alongside the Bulu' Sipong mound and increased the number of water trucks that wet the now-sealed road to reduce the amount of dust produced by heavy vehicles driving by. 'Probably the most reassuring development is that [Semen Tonasa now] has a dedicated sustainability team working on this issue,' says Whincop.

Lebe and his team at the cultural heritage office lack the resources to fund sophisticated investigations into the plethora of variables that affect cave art deterioration. In that respect, Peterson says, it might turn out to be fortunate that Bulu' Sipong 4 lies in Semen Tonasa's mining concession. The company has the means to fund a comprehensive monitoring program that could end up providing crucial clues about what's harming caves throughout the Maros–Pangkep karst landscape, says Peterson.

Along with the shortage in resources, there is also limited awareness of the caves and their cultural heritage. 'As an Indonesian, I only just found out that there are more than 500 caves in the Maros–Pangkep area, which is astonishing,' says Daunan. 'Not that many Indonesians know about this,' she says.

That could be changing slowly. In May, the karst landscape stretching across Maros–Pangkep was officially declared a UNESCO Global Geopark. This status could attract extra resources from the Indonesian government in the future. But Huntley says that more urgent investigations to stem the deterioration are needed. Meanwhile, Brumm is trying to create a digital record of the cave art, which could help with its conservation and enable people to experience the paintings without harming them. 'What we do need to do right now,' he says, 'is to create a 3D visual archive of all of the rock art there, before it's gone.'

Standing in the cave next to the peeling paintings, it's hard to grasp their vast age. It's equally hard to fathom how many more hand stencils, warty pigs and hunting scenes might be tucked away in as-yet-undiscovered caves in Maros–Pangkep – and that they might disappear soon, never revealing their magnificence to modern eyes.

❋ *Satellite tracking the Pacific's most endangered leatherback turtles*, p. **14**
This little theory went to market, p. **135**

PREDICTING THE FUTURE

Drew Rooke

The computer screen radiates line after line of strange text: black letters, numbers and symbols on a light grey background. Scrolling through, it seems to be never-ending and – at least for me – mostly unintelligible; I'm not fluent in Fortran, the programming language it's written in. But there's a plain English preface to this gobbledegook text.

'Many people have contributed to the development of this code,' it reads. 'This is a collective effort and although individual contributions are appreciated, individual authorship is not indicated.'

I'm visiting the Climate Change Research Centre (CCRC) at the University of New South Wales (UNSW), Sydney, and what I'm looking at is the backend of a modern-day climate model – specifically, the University of Victoria Earth System Climate Model (UVic ESCM).

This is just one of many models that scientists use nowadays to peer into Earth's past and make projections about its future. For example: that our planet is likely to warm 1.5 degrees Celsius above pre-industrial levels in the 'near-term' and that 'every increment of global warming will intensify multiple and concurrent hazards', as the IPCC Sixth Assessment report says.

Like many people, I strongly believe these projections. But my belief is based on little more than blind faith in the models that produce them. In fact, I know so little of the inner workings of climate modelling that it seems to me like something of a dark art

– a form of planetary-scale fortune telling. The reason I'm here is to break the spell; to see for myself exactly what is behind the curtain of a contemporary climate model.

My guide is Katrin Meissner, director of the CCRC and a professor at UNSW, who melds a brown bob, warm smile and more than 25 years' experience in climate modelling. She first started developing models while completing her PhD in the mid-1990s at the Alfred Wegener Institute for Polar and Marine Research in Germany, and initially 'hated' the computer coding involved. Over time, it grew on her and she came to appreciate its beauty.

'You have to use your brain to try to work out why things don't work; it's like a puzzle you have to solve,' she says.

In 2000, Meissner accepted a postdoc position in the School of Earth and Ocean Sciences at the University of Victoria, Canada, where her coding experience and knowledge of the Earth's climate system combined to help her build the very model I'm looking at now.

In climatologist parlance, the UVic ESCEM is a model of 'intermediate complexity': it couples key processes that influence the Earth's climate system – including the atmosphere, ocean, sea ice and land surface – but has a low resolution in order to run long-term simulations of large-scale processes that allows its driving computer hardware to run efficiently. Other types of climate models include more basic zero-dimensional 'energy balance models', which have a very coarse resolution and only simulate incoming and outgoing radiation at the Earth's surface. Much more sophisticated 'general circulation models' have a much finer resolution and simulate the whole climate system on a global scale in three dimensions.

Meissner describes the use of any one of these models as 'a very dry scientific approach' to studying the Earth's climate, which mirrors the mindset of their makers – a point often overlooked by their detractors. 'We are actual physicists and mathematicians who just want to understand the science.'

Earth capture

Earth's climate is best described as chaos: a system that's highly responsive to the smallest perturbations – or, to speak more scientifically, that exhibits 'a sensitive dependence on initial conditions'. As pioneering meteorologist Edward Lorenz, who made famous this profound phenomenon, said in 1972, it's possible that even the flap of a single butterfly's wings in one location can create a positive feedback loop that ultimately leads to a tornado forming faraway in the future – or 'can equally well be instrumental in preventing a tornado'.

But even though this system is inherently distempered, it is also governed by fundamental physical laws, such as the conservation of mass, energy and momentum – which is why Gavin Schmidt, director of the NASA Goddard Institute for Space Studies (GISS), says that developing a virtual representation of it is 'Quite straightforward. Well, compared to general relativity, it's straightforward.'

Long before he became a world-leading climate scientist, Schmidt 'bummed around' in Australia for a year, 'working in restaurants, picking grapes, running a youth hostel in Perth. It was a lot of fun, but I wasn't doing anything terribly intellectual and I got a little bored.'

He cut his modelling teeth shortly after he completed his PhD in applied mathematics in 1994 at University College London. 'A friend of my supervisor was looking for people researching climate, constructing models of the thermohaline circulation in the ocean. And I said, "Oh, that's absolutely fascinating." But I had no clue what that meant.

'I remember putting the phone down and then asking my supervisor, "What the hell is the thermohaline circulation?"'

Lack of knowledge managed, Schmidt took the job at McGill University in Canada, which turned into 'a big crash course in climate writ large'. Then, in 1996, he moved to New York – 'because I had a girlfriend there' – and began at GISS, developing climate models. These models are based on the same laws of physics as govern Earth. But in models, physics laws are represented mathematically by the 'primitive equations' first formulated by Norwegian physicist and meteorologist Vilhelm Bjerknes at the turn of the 20th century, and later encapsulated in code.

Typically containing enough code to fill 18 000 pages, modern global climate models are actually a composite of separate models of individual components of Earth's climate, such as the atmosphere, ocean, land surface and sea ice. These are carefully combined – an intricate process that Simon Marsland, the leader of the National Environmental Science Program Climate Systems Hub, likens to 'doing brain surgery on many different brains at the same time and connecting wires in between them'.

They work by dividing the entire globe into a three-dimensional grid of thousands of interconnected cells. 'Kind of like a chess board,' says Marsland, who is based at the CSIRO.

This gigantic grid's individual cells represent segments of the atmosphere, ocean, land surface or cryosphere. Their size defines a model's spatial resolution and therefore its computational cost. Nowadays, they are typically 50–200 square kilometres in size and up to 90 layers tall in the atmosphere and 60 in the ocean.

Scientists 'initialise' each cell – that is, they encapsulate in code its initial temperature, air pressure, humidity and other important physical processes, using the trove of observational climate data gathered over the years by ships, satellites and atmospheric balloons. But this is an imperfect art; there are many key processes that are either not fully understood or act on subgrid scales, such as the formation of clouds, which have to therefore be approximated.

After 'external forcings' – like solar radiation and greenhouse gas emissions – have been included, the model can then begin the protracted process of calculating the huge number of equations coded into every cell at discrete time intervals, typically one hour.

Inevitably, errors appear. But once the model has been carefully refined and 'spun-up' – that is, run over an extended period of time – something remarkable happens. A realistic, albeit simplified, simulation of the Earth emerges, complete with what Schmidt calls 'emergent properties' that haven't even been explicitly coded, such as the Hadley circulation (air rising near the equator and flowing polewards) or large storms in the mid latitudes.

'We're not coding those properties; they just emerge out of the interaction of the individual cells,' says Schmidt. 'Yet they match what the real world does. And you go, "Well, that's incredible! Who would have guessed that this would have worked?"'

A model needs to reach this state of equilibrium for scientists to have confidence to start using it to simulate, for example, the long-term consequences of a huge volcanic eruption, or a dramatic spike in the level of carbon dioxide in the atmosphere.

But in order to account for the inherent chaos in Earth's climate system, they run multiple simulations with slight changes in the initial conditions to get a comprehensive understanding of the probability of different outcomes occurring. This process, in the case of the most sophisticated models looking far into the future, can take up to a year to complete on the most powerful supercomputer. Meissner best sums up these models: 'They're beasts.'

Laying the foundations

Climate-modelling beasts are the culmination of more than 100 years of evolution, the roots of which tap into the work of Swedish chemist Svante Arrhenius.

In 1896 – seven years before he won the Nobel Prize in Chemistry for his work on the conductivities of electrolytes – Arrhenius published the seminal climate science paper 'On the influence of carbonic acid in the air upon the temperature of the ground'.

It was well known by then that the carbon dioxide – or, as it was more commonly known at the time, carbonic acid – in Earth's atmosphere trapped incoming solar radiation and thus had a potent warming effect on its surface temperature.

This had first been demonstrated by an unpaid American scientist named Eunice Foote in 1856; it took more than 150 years for the scientific community to recognise her achievement.

In a series of simple experiments at her home in Seneca Falls, New York, Foote measured the temperature of small glass cylinders filled with different gases and placed in the sun and shade. Most notably, she found that the cylinders containing carbon dioxide not only warmed the most in sunlight, but also took the longest to cool on being removed from it.

As she wrote in a short paper titled 'Circumstances affecting the heat of sun's rays': 'An atmosphere of that gas would give to our Earth a high temperature; and if as some suppose, at one period of its history the air had mixed with it a larger proportion than at present, an increased temperature ... must have necessarily resulted.'

Arrhenius wanted to find out just how much warmer – or colder – the Earth would be if the proportion of carbon dioxide in its atmosphere drastically changed. In his paper he presented the results of a very simple theoretical model he had used to come up with an answer.

His model divided the Earth into 13 latitudinal sections from 70°N to 60°S. Armed with only pencil and paper and grieving the end of a brief marriage with one of his former students, Arrhenius then calculated for each section the temperature under six different carbon dioxide scenarios, using the best available information about how much incoming solar radiation was absorbed by atmospheric carbon dioxide and water vapour.

These 'tedious calculations', as he described them, took him the better part of a year to complete. As well as demonstrating that a significant reduction in atmospheric carbon dioxide could trigger an ice age, they showed that a doubling of this potent greenhouse gas would warm Earth's average temperature by 5–6 degrees Celsius – an estimate he hoped would materialise. As he told the crowd at a lecture at Stockholm's Högskola (now Stockholm University) in 1896, global warming would permit 'our descendants ... to live under a warmer sky and in a less harsh environment than we were granted'.

Arrhenius's paper was the first to quantify the contribution of carbon dioxide to the greenhouse effect – and even though it was crude, it laid the foundations for much of the climate modelling work that followed. As NASA's Gavin Schmidt says: 'Our understanding of the climate is subtler now. We have more data. But scientists at the turn of the 20th century – they were mostly correct about how things worked and how things would change. They were extremely insightful. They had it all worked out.'

Time to complexify

A key step in the evolution of climate models occurred in 1922, when British mathematician and meteorologist Lewis Fry Richardson published *Weather Prediction by Numerical Process*. In this now-classic book, Richardson described a radical new scheme for weather forecasting: divide the globe into a three-dimensional grid and for each grid cell, apply simplified versions of the primitive equations to deduce its weather.

This was a far more scientific method of forecasting the weather than looking for similar patterns in records and extrapolating forwards, as was conventionally done, but it also consumed much more time. In fact, it took Richardson six weeks performing calculations by hand just to produce an eight-hour forecast for one location (which, due to issues with his data, was wildly inaccurate).

Richardson imagined a fantastical solution to this problem: a 'forecast factory'. A large circular theatre, it would have a gridded map of the world painted on its wall. Inside, 64 000 mathematicians – or human computers – each armed with a calculator would solve the equations in their assigned cell and develop a weather forecast in real time. In the centre of the factory, the man in charge of the whole operation would sit on a large pulpit, coordinating the human computers like an orchestra conductor with coloured signal lights.

Richardson's fantasy became something of a reality in 1950, when a team at Princeton University in the US adopted his grid-cell scheme and produced the world's first computerised, regional weather forecast using the first general-purpose electronic computer, known as ENIAC. But scientists knew that the same scheme could be used to construct a three-dimensional model that simulated not just regional weather, but the global climate as well.

With this goal in mind, the United States Weather Bureau established the General Circulation Research Section in 1955, under the direction of US meteorologist Joseph Smagorinsky. Fresh from working as a weather observer during World War II, Smagorinsky was eager to recruit the finest scientific minds from around the world.

There was one young scientist from Japan in whom he sensed prodigious potential: Syukuro Manabe. Born in 1931 on the island of Shikoku, Manabe had survived his wartime upbringing to become one of the leading students of meteorological physics at the University of Tokyo – and upon joining Smagorinsky's team in 1958, he was tasked with managing the completion of its establishment goals.

Manabe knew this wouldn't be easy, but he had a powerful new tool in his kit: an IBM 7030, also known as 'Stretch' – the world's first supercomputer. Larger than a small house and weighing roughly 35 tonnes, it had initially been developed (at the Pentagon's request) for modelling the effects of nuclear weaponry. It could perform more than 100 billion equations in about a day, but this wasn't enough computing power to run a three-dimensional climate model.

So with his colleague Richard Wetherald, Manabe built a simplified one-dimensional version that incorporated key physical processes such as convection and radiation to simulate the transfer of heat throughout a single column of air 40 kilometres high in a number of different carbon dioxide scenarios.

In May 1967 – by which time the General Circulation Research Section had been renamed the Geophysical Fluid Dynamics Laboratory (GFDL) – Manabe and Wetherald published their findings. Their paper 'Thermal equilibrium of the atmosphere with a given distribution of relative humidity' didn't sound particularly exciting. But it proved to be a game changer; in fact, in a 2015 survey, the world's leading climate scientists voted it as the most influential climate science paper of all time.

There are numerous reasons. For one, the paper marked the first time that the fundamental elements of Earth's climate had been represented in a computer model. But it also provided the first reliable prediction of what doubling carbon dioxide would do to overall global temperature. As Manabe and Wetherald wrote: 'According to our estimate, a doubling of the CO_2 content in the atmosphere has the effect of raising the temperature of the atmosphere (whose relative humidity is fixed) by about 2°C.'

Manabe would ultimately go on to win the Nobel Prize in Physics in 2021 for his contribution to the physical modelling of Earth's climate. In 1969, he and Kirk Bryan – an oceanographer who also worked at the GFDL – produced the first ever coupled climate model. It combined Manabe and Wetherald's famous atmosphere model with one Bryan had previously developed of the ocean, and although it was highly simplified and took roughly 1200 hours to run on a UNIVAC 1108 computer, it successfully produced many realistic features of Earth's climate.

By 1975, computing technology advanced enough for Manabe and Bryan to construct a more detailed version of their coupled model prototype. With a grid size of 500 square kilometres, it included 'realistic rather than idealised topography' and successfully simulated 'some of the basic features of the actual climate'. But it also simulated many 'unrealistic features', underscoring the necessity of increasing climate model resolution in the future.

This became possible towards the end of the 20th century. Computer processing power dramatically increased, enabling them to complete larger and larger calculations. By 1995, for example, the typical horizontal resolution of global climate models was 250 kilometres; six years later, it was 180 kilometres.

At the same time scientists were increasing climate model resolution they were adding in more and more important components. By the mid 2000s, some models were integrating not only the atmosphere, land surface, ocean and sea ice, but also aerosols, land ice and the carbon cycle.

But even though climate models have become infinitely more complex over the decades, the physics and mathematics at their core has, as the CSIRO's Simon Marsland says, 'essentially remained the same'.

The proof is in the volcano

On 15 June 1991, after months of intensifying seismic activity, Mount Pinatubo – located on Luzon, the Philippines' largest and most populous island – exploded. As rocks and boiling mudflows surged down the volcano's flanks, a toxic ash cloud soared roughly 40 kilometres into the sky and, over time, completed several laps around the globe.

The eruption – the second largest of the 20th century – killed nearly 900 people and left roughly 10 000 homeless. It also released 17 megatons of sulphur dioxide into the atmosphere, which drew the attention of climate scientists.

It had long been known that volcanic aerosols reflect sunlight back to space and thus temporarily reduce Earth's temperature, and almost immediately after the eruption, climate scientist James Hansen and several of his colleagues at NASA's GISS started running experiments with their climate model to estimate how much – and for how long – it would cool the planet.

In January 1992, they published their findings. The eruption, they wrote, was 'an acid test of climate models' as the 'global shield of stratospheric aerosols' would have a strong enough cooling effect that could easily be simulated.

According to their model, 'intense aerosol cooling' would begin late in 1991 and maximise late in 1992. By mid 1992, it would be large enough to 'even overwhelm global warming associated with an El Niño that appears to be developing'. Critically, their model also showed that by the later 1990s, the cooling effect would wear off and there would be 'a return to record warm levels'.

As it turned out, this is exactly what happened. But this is far from the only evidence of the reliability of climate models.

In 2019, a team led by climate scientist Zeke Hausfather published an extensive performance evaluation of the climate models used between 1970 and 2007 to project future global mean surface temperature changes. After comparing predictions with observational records, they found that 'climate models published over the past five decades were generally quite accurate in predicting global warming in the years after publication, particularly when accounting for differences between modelled and actual changes in atmospheric CO_2 and other climate drivers'.

The researchers added: 'We find no evidence that the climate models evaluated in this paper have systematically overestimated or underestimated warming over their project period. The projection skill of the 1970s models is particularly impressive given the limited observational evidence of warming at the time.'

In fact, one such model, developed by Manabe and Wetherald, not only produced a robust estimate of global warming, it also accurately forecast how increased carbon dioxide would affect the distribution of temperature in the atmosphere. As they wrote in a 1975 paper describing the model: 'It is shown that the CO_2 increase raises the temperature of the model troposphere, whereas it lowers that of the model stratosphere.'

Stratospheric cooling is now recognised as the fingerprint of global warming. But as Schmidt explains: 'There was no reason to think that was right other than the mathematics suggested it. Remember, this was before anybody had any temperature measurements in the stratosphere, let alone trends.

'But now we've had 40-odd years of satellite information from the stratosphere and troposphere and surface, and we can see that what's happened is more or less exactly what was predicted.'

This is the basis of Schmidt's confidence in these models: 'that from the beginning, they have made predictions that have come true.'

But climate models – even the most advanced ones – do have limitations. Many important small-scale processes are not accurately represented in them, due to ongoing computational constraints which prevent finer resolutions. One such process is the formation of clouds, which play a key role in regulating the planet's temperature.

Models are also inevitably biased – they might be either slightly too cold or too warm, for example. Over time, these biases, no matter how small, will amplify and can cause a model to drastically move away from its initial conditions – a phenomenon referred to as 'drift'.

But even as computers become more powerful, allowing for finer resolution, and scientists reduce the biases and drift in models, there is one limitation that can never be overcome. The Earth remains inherently chaotic, containing an infinite multitude of interacting conditions that range from the microscopic to planetary in size; it's impossible to build a perfect replica of it.

'We're never going to have a perfect model; our models are always going to be wrong,' Schmidt says, echoing every one of his peers interviewed for this story. 'But they are demonstrably useful. It's kind of incredible: even though they are imperfect, incomplete and still have bugs in them, they nonetheless make useful, good predictions. And those predictions have been getting better over time.'

And even though the work that's entailed in both developing and using these models can be, according to Marsland, 'very slow', 'very frustrating', and 'prone to many failures', it has also been unpredictably life-changing for him.

'I think the more years you work in this field, the more you get exposed to processes happening in the world that you would have no idea about otherwise. Your eyes are continually being opened and you're always being reminded that the world is a beautiful place and that we have a responsibility to make sure it remains a beautiful place for the people who come after us.'

✱ *Western Australia had its hottest summer ever, but climate change barely made the news*, p. **22**
   Science in the balance, p. **208**

THIS LITTLE THEORY WENT TO MARKET

Elizabeth Finkel

Reporting on the origins of the Covid pandemic has been a challenge for journalists. From the beginning there have been two narratives. One was that the virus, SARS-CoV-2 (SARS-2), jumped from an infected animal into humans at the Huanan Seafood Market, in the city of Wuhan, China. The other was that the virus was genetically engineered at one of the two campuses of the Wuhan Institute of Virology, which lie 12 and 27 kilometres to the south of the market across the Yangtze River. After accidentally infecting a lab worker, it spread into Wuhan and ultimately the world.

In short: natural origin versus a lab leak.

Some journalists became proponents of the lab leak and were taunted as 'conspiracy theorists'; others followed the mainstream scientific view and were taunted as 'stenographers of science'.

My inclination was natural origins, given that the first SARS outbreak had arisen this way, as had Ebola, Middle East respiratory syndrome (MERS), HIV, bird flu outbreaks and others. But the accusation of 'playing stenographer' stung. It forced me to check my process. I do not blindly trust scientists. My lodestone is the scientific method itself.

It works like this. You form a hypothesis. You gather data to test it. That doesn't mean only trying to validate it, it also means trying to see if you can disprove it. Most of the time, science can't prove things; it advances based on the 'weight of evidence'.

What I want to do here is show how I weigh the evidence based on the best two sources I can muster: the peer-reviewed scientific literature and, in a novel experience for me, the United States intelligence agencies that reported their findings to the Biden administration in June.

The bottom line: the evidence weighs heavily on the scale towards natural origins, leaving the lab leak side a lightweight proposition.

Before I take you through the evidence, it's instructive to revisit the first SARS outbreak beginning at the end of 2002. It claimed a few lives in Europe and the Americas, taking its major toll in Hong Kong and China: at least 770 deaths. It seems small fry by comparison with SARS-2 that claimed at least six million. But SARS-1 was a wake-up call. Till then, coronaviruses seemed relatively harmless, agents of maladies such as the common cold. After SARS-1, the international research community mobilised to try and stop such a thing from happening again.

The key was to figure out where SARS-1 had come from. The first part of the answer was easy: palm civets, weasel-like animals, sold at wildlife markets in Guangdong. Here, live animals, often trapped in the wild but nominally produced by the wildlife farming industry, were butchered and consumed at nearby eateries. But civets could not provide the whole story. No significant wild or farmed population outside of the markets seemed to carry the virus, so its evolutionary source likely lay elsewhere. For virologist Linfa Wang, formerly at CSIRO, now at Duke-NUS Medical School in Singapore, it was most likely a bat. Unlike other mammals, bats show remarkable tolerance to coronaviruses, allowing them to proliferate without themselves sickening. Collaborating with Zhengli Shi's lab at the Wuhan Institute of Virology (WIV), researchers searched bat caves for SARS-like viruses. Some of the research was funded by the not-for-profit US-based EcoHealth Alliance headed by Peter Daszak.

In 2013, the researchers found the source in bat caves in Yunnan province in China's south-west: coronaviruses that were 95 per cent similar to SARS-1. They also found something alarming. The viruses were evolving fast, producing weird new combinations. Tests at WIV and at Ralph Baric's lab at the University of North Carolina showed these novel combinations were able to infect human cells.

Warnings were issued. The EcoHealth Alliance produced pamphlets to educate villagers on how to live safely around bats. And in international publications from 2013 to 2017, the international team – Shi in China, Baric and Daszak in the United States, and Wang in Singapore – warned of the clear and present danger.

In December 2019, SARS-2 emerged in Wuhan.

Over half of the initial cases were people who worked at the Huanan Seafood Market or had been in contact with someone who had. On 1 January 2020, the market was shut and disinfected. By 24 February, a national law was put in place outlawing the eating of wild animals and the wildlife trade that supplied them. It seems China was quite clear about the origins of SARS-2: it had jumped ship from a wild animal at the Huanan Seafood Market.

Yet by February 2020, the world was already entertaining another hypothesis: the lab leak.

Ironically, the two researchers who first raised the possibility now find themselves accused of trying to cover it up.

Eddie Holmes, at the University of Sydney, and Kristian Andersen, at Scripps Research in San Diego, are leading lights in the field of virus evolution. In August, the Royal Society in the United Kingdom recognised Holmes's contribution with the Croonian Medal, the same prize Howard Florey received for developing penicillin. These two boffins spend their waking hours poring over the family trees of viruses.

In mid January 2020, the pair took their first look at the 30000-letter genetic code of the SARS-2 virus – and got a shock.

There were two strange signatures not seen in SARS-1. Holmes was alarmed by a 20-letter string of code lying within the corona-virus's homing device, known as the receptor-binding domain. Like SARS-1, it targeted a receptor on human lung cells, but with even better precision. Andersen was alarmed by a 12-letter string of code that was not present in SARS-1 at all. Known as a furin cleavage site, it helped unleash the homing device. Both new signatures might explain the super-spreading power of this new virus.

Could the signatures have arisen from genetic engineering experiments at the WIV? The Wuhan institute was a world-renowned sentinel for monitoring bat viruses. Its freezers contained thousands of bat viruses and it conducted experiments to test their function. Could such a test have gone awry?

Their concerns quickly reached Anthony Fauci, then head of the US National Institute of Allergy and Infectious Diseases (NIAID) and stoic adviser to President Donald Trump. Fauci recommended two things: that a conference call be conducted including some of the leading experts on virus evolution and public health, and that they contact their respective intelligence agencies to share their concerns. In Australia, Holmes contacted Nick Warner, then head of Australia's Office of National Intelligence.

The experts got together on February 1. They brought new data mined from virus databases. During the course of that meeting, which included a mostly silent Fauci, and over the next few days, Holmes and Andersen changed their minds – they concluded that SARS-2 was not likely to be the result of a lab leak. Exactly what scientists are supposed to do when faced with new data.

But that change of mind gave birth to a conspiracy theory. Journalists such as Australia's Sharri Markson join the dots as follows: Fauci's agency NIAID had funded work at the WIV. So Fauci was trying to cover up the evidence of the lab leak to prevent reputational damage. Andersen was the recipient of part of a US$9 million grant from NIAID, so he was motivated to collude with Fauci. And Holmes? Markson has repeatedly accused him of being part of the cover-up, but it's not clear what motive he's supposed to have had.

None of this constitutes evidence. They are *ad hominem* attacks. But for the sake of the exercise, let's test how plausible these assertions are. Fauci claims he had no idea of the NIAID funding to the Wuhan institute, pointing out that his department funded projects to the tune of US$6 billion per year. The institute received US$120 000 per year, subcontracted via the EcoHealth Alliance, an amount hardly likely to come to his notice.

Second, the claim that Andersen was colluding with Fauci to protect his grant? Indeed, a $9 million grant would have been more noticeable, but NIAID grants are not awarded by fiat of Fauci. There is an independent peer-reviewed process. Andersen's grant had scored winning marks a month before the February 1 meeting.

And, while we're dispensing with the lightweight stuff, let's include another bit of 'evidence' typically offered in favour of a lab leak: the claim that the first Covid-19 patient in Wuhan was a researcher at the WIV. US intelligence agencies looked for some substantiation of that and found none.

What was the evidence that persuaded Holmes and Andersen to change their minds in February 2020?

After searching the databases, it turned out that the two alarming signatures in the genetic code of SARS-2 were not unusual for coronaviruses. Regarding the furin cleavage site, MERS had one, and so did the common cold coronaviruses. The receptor binding domain in the virus homing device was not new either. It turned up in a coronavirus that had been isolated from illegally traded pangolins. If nature was cooking up these code changes in coronaviruses, there was no longer any compelling reason to propose a man-made origin.

A second line of evidence against a laboratory-made virus came from considering how plausible it would have been to create it. For starters, if genetic tinkering at WIV had produced SARS-2, then what starter virus was used as its chassis? While the institute had thousands of bat viruses in its freezers, most were not intact viruses but fragments extracted from bat poo. Only three viruses were intact enough to be grown in animal cells. One was a SARS-1 ancestor, called WIV-1. But it could not have served as the chassis for SARS-2: it was only 79 per cent similar across the 30 000 letters of code.

Zhengli Shi, frantic that her lab might be responsible for an accidental leak, examined all the genetic codes that had been read for the bat coronavirus in her freezers. One gave her pause. Isolated in 2013, its genetic code was a 96 per cent match to SARS-2. But the virus, named RaTG13, was one of those Shi had never been able to grow in animal cells. If she had, we'd have heard about it. Her stellar CV included regular publications on newly discovered viruses in *Nature* and *Science*.

Armed with that evidence, Holmes and Andersen changed their thinking over a few days in February. The balance of probabilities showed that this new virus was not made in the laboratory of man, but in nature.

In the three and a half years since, the evidence has mounted in favour of that assessment. We'll get to it in a moment.

But first, we need to examine a more substantive piece of evidence for the lab leak hypothesis.

In September 2021, a group of lab leak sleuths unearthed a failed grant proposal titled 'DEFUSE'. Submitted by the EcoHealth Alliance in 2018 to the US Defense Advanced Research Projects Agency (DARPA), its ultimate goal was to 'defuse' the risk of virus spillovers from bats. Part of that work involved taking bat viruses and engineering different furin cleavage sites into them to test their effects. The work was to be a collaboration between the WIV, which would supply the bat viruses, and Ralph Baric's group at the University of North Carolina, who would do the genetic engineering. DARPA rejected the grant proposal. Baric says the work did not go ahead.

Nevertheless, reports of the DEFUSE grant hit like a bombshell.

The lab-leakers saw it as the smoking gun that explained how SARS-2 acquired its extraordinary furin cleavage site. Even if DARPA failed to fund the project, the plans were in place. It might have gone ahead.

But evidence that it happened? Two years on there is none.

Some hypothesise that the DEFUSE work took place at the WIV and the virus leaked, because work there was undertaken at a moderate biosecurity level, rather than the higher biosecurity level that would have been used in North Carolina.

But that argument still fails the test of what starter virus was used. Even if researchers had been able to find a way to grow the close relative of SARS-2 – RaTG13 – there are over 1000 letters of genetic code that are different. Experts such as Stuart Turville at the University of New South Wales, who engineers new viruses for gene therapy, says the odds of being able to artificially introduce such huge chunks of code are extremely low. Wang puts it more bluntly. Until nature had showed us what it took, he tells me, 'no scientist, smart or rich, would be able to make SARS-CoV-2'.

And even if you believe, as the lab leak champions do, that Shi is covering up – she has, for instance, declined to surrender her WIV lab books – that is not true for Baric. The intelligence agencies would have had free access to his lab. Their finding was that no chassis existed.

Hence, an evidence-based approach leads to dismissal of the DEFUSE grant argument.

Finally, let's get to the scientific evidence that has emerged over the past three years that weighs heavily in favour of a natural origin for SARS-2.

First, the hunt for the bat ancestors is proceeding much like the hunt for the SARS-1 ancestor did. Researchers then never found a perfect match – the closest, YN2020E, was 96 per cent similar. Rather, they found several relatives, each of which carried a different part of the SARS-1 virus. The most likely explanation is the coronaviruses swapped bits to produce the ancestor of SARS-1.

So far, the hunt for SARS-2 ancestors has likewise delivered several family members. French researchers found one group of viruses in bat caves in Laos. Dubbed BANAL, they showed a 96.8 per cent match overall, and almost a 100 per cent match over the virus homing device, the receptor binding domain. Another bat coronavirus found in Yunnan province, RmYN02, showed a lower 93 per cent match overall, but a 97 per cent match across a SARS-2 gene called '1ab'. Examining the genetic codes of these relatives, Jonathan Pekar, a bioinformatics expert at the University of California San Diego, calculates the SARS-2 ancestor may have circulated as little as three years prior to the outbreak.

None of the bat coronaviruses so far appear to show a match to the furin cleavage site. Pekar believes the most likely scenario is that this bit of code evolved in an intermediate animal host, perhaps a palm civet or raccoon dog – animals susceptible to SARS-2 infection and known to be sold at the Huanan Seafood Market.

We may never discover the original host. Whatever infection the species carried has likely been swamped by the SARS-2 that jumped from humans to animals – it has become rife in populations from American white-tailed deer to Dutch mink.

But the major animal host for SARS-2 – humans – reveals something interesting. As SARS-2 has evolved from the original Wuhan strain to Delta, Omicron and now Pirola variants, the furin cleavage site keeps changing. For Gary Whittaker, a Cornell University virologist who tracks the evolution of this chunk of code, it's evidence that the furin cleavage site is 'the design of nature, not man'.

A completely different line of evidence comes from evolutionary biologist Michael Worobey at the University of Arizona. Worobey has a reputation for trying to 'disprove' theories. Two decades ago, he travelled to the Democratic Republic of Congo to disprove the dominant theory that HIV crossed over from wild chimps – the alternative being that it crossed over from a contaminated polio vaccine. His findings proved the dominant theory correct.

With the outbreak of SARS-2, Worobey was suspicious about reports linking the earliest cases to the Huanan Seafood Market. Perhaps an unseen factor was skewing the results? For instance, market workers might have been the first to be tested. To disprove the market origin hypothesis, he mapped the home addresses of the first 150 cases onto a map of Wuhan, a sprawling megacity of 11 million. The addresses clustered around one spot: the seafood market. 'Long story short, they all lived absurdly close to and centred on the wildlife market; the highest probability of drawing a case is within a bubble a 100 metres from the market,' Worobey told me. His paper providing these findings was published in *Science* in 2022.

But was the market simply the site of a super-spreading event? If so, then it should have spread a single strain of the virus, but another paper in *Science* in 2022, by Jonathan Pekar, showed the market harboured not just one strain, but two. That's far more consistent with what is known to happen with animal viruses, which make hopeful leaps to humans all the time.

China has been completely unhelpful about resolving the origins debate. Its authorities declined to hand over researchers' lab books, but they also declined to share crucial data from the Huanan market. For instance, we now know that swabs taken from the market prior to the shutdown contained DNA from several wild animal species susceptible to SARS-2, including raccoon dogs and civets. Many of the swabs also contained SARS-2 – implicating one of these species as the original carrier. Yet it took more than three years for this evidence to be made publicly available.

Why has China been so unhelpful? Competing interests, says Peter Li, a wildlife policy scholar based at the University of Houston-Downtown. The wildlife trade – once a US$74 billion industry – is fighting for survival. Li says the lab leak theory has helped stage a 'spirited comeback' because it exonerates the industry from blame.

Li traces the beginning of China's stonewalling to May 2020, when Trump's trade adviser Peter Navarro accused China of seeding Covid-19 by sending 'hundreds of thousands of Chinese on aircraft to Milan, New York and around the world'.

Toxic politics from both sides have damaged science. The vibrant collaborations between Chinese and Western scientists that constituted an early warning system for emerging viruses have been severed. And in their own countries, Holmes, Andersen and colleagues are sapped of energy as they fend off media smear campaigns and the harassment and death threats that follow. We are arguably in a far worse state of readiness for the next inevitable pandemic.

The media bear much of the responsibility. As the scientific evidence mounts in support of a natural origin, public opinion veers in the opposite direction. Polls show that some two-thirds of Americans believe the lab leak hypothesis is correct. Journalists need to examine their process. Rather than wedding themselves to a particular narrative, they can do no better than taking a page from the scientific method.

✱ *In the heart of the forest, one woman built a house of slime*, p. **8**
Call of the liar, p. **198**

HOW SCIENTISTS SOLVED THE 80-YEAR-OLD MYSTERY OF A FLESH-EATING ULCER

Angus Dalton

The sore on David McLachlan's ankle swelled to the size of a saucer. For Alice Mika, another Melbourne resident, a small bump grew into a dollar-sized ulcer and eluded diagnosis for months. And for retiree Fiona Wordie, as the wound on her foot expanded, so did her pet cockatoo's vocabulary; the ulcer caused such regular pain the bird learned to squawk the f-word.

Such is the misery wreaked by the Buruli ulcer, a flesh-eating tropical oddity that began to afflict residents near Melbourne more than 80 years ago. But in January 2024, the publication of a *Nature Microbiology* paper marked the end to the quest to discover how this terrible disease spreads.

And it couldn't have come too soon. Cases of the ulcers are increasing year-on-year. In 2021, the flesh-eating ulcer spread from the Mornington and Bellarine peninsulas to inner Melbourne and, in September 2023, a case was identified in Batemans Bay, marking the leap of the disease to southern NSW.

Researchers have long suspected mosquitoes carried the ulcer-causing bacteria from possums to humans, causing a festering wound to bloom where the blood-sucking insects had sunk in their proboscis.

A small, dogged group of scientists spent almost three decades building a hierarchy of evidence to convince themselves, their colleagues and health departments of the veracity of their hypothesis. This is how they did it.

A stealthy, lab-resistant, skin-melting bug

The bacterium that causes the ulcers, *Mycobacterium ulcerans*, is devious. Australian scientists first described the bacteria in 1948, but since then the bug has evaded easy inquiry.

It can't be cultured in a petri dish. It's extremely slow-growing. In the body, the bacterium cloaks itself in a lipid called mycolactone to conceal itself from the immune system. The toxin it emits melts skin and fat. After a few weeks of infection, the wound collapses and the gnarly face of the flesh-eating ulcer is revealed.

Tim Stinear, now a professor at the Doherty Institute and head of the World Health Organisation lab tackling *M. ulcerans*, was a young microbiologist working in water testing for the Victorian government in the mid-'90s when the ulcers first came to his attention.

'I got a call one day from a clinician who was investigating a mysterious outbreak of flesh-eating ulcers on Phillip Island,' Stinear says. 'He'd heard from the health department that I had this newfangled molecular test that could detect pathogens in the environment.'

The caller was Professor Paul Johnson, who would also play a central role over the next three decades in untangling the mystery. He wanted to use a new kind of PCR test Stinear could access to hunt the offending bacteria.

At this stage researchers knew the ulcers didn't spread from person to person, and probably came from the environment. They used the tests Johnson had called Stinear about to scour golf courses, soil and water near infected people. No dice.

But then, in 2002, Johnson tested leaves, soil, bugs and pellets of possum poo from Point Lonsdale. The possum scat returned a result 10 000 times stronger than anything they'd seen. Researchers searched the area's trees with torches and spotted possums afflicted by ulcers. They lured 42 possums into traps with apples and peanut butter. A quarter of them carried the bacteria.

Johnson also noticed that the placement of the ulcers in humans was odd. His patients had the sores on the tips of their ears, their backs and legs – spots where mosquitoes typically fed. In 2007, the investigators reported mosquitoes trapped in Point Lonsdale also carried *M. ulcerans*. A compelling story had emerged.

Building the hierarchy

No one was particularly convinced.

'We'd present our data in public meetings, in WHO meetings in Geneva. People were interested but they weren't compelled by the data,' Stinear recalls.

'They said "Oh, well, it could be some other biting insect, or maybe the insects are contaminated with the bacteria, but they're not actually spreading it."'

In 2019, scientists were still publishing reports that cast doubt on the mosquito hypothesis. There were also no examples of mosquitos transmitting a mycobacterial infection of the type that causes leprosy, tuberculosis and Buruli ulcers. 'We were going against any precedent,' Stinear says.

The scientists set their sights on a more forensic approach that would convince the toughest reviewers.

Over five years, the researchers, led by Agriculture Victoria's Dr Peter Mee, caught 73 580 mosquitoes. A handful were fat with blood.

The scientists analysed the blood in the engorged mosquitoes. From 36 blood-fed mosquitoes, 20 had fed on ringtail possums. And, crucially, three had feasted on possums and humans. Bingo.

'It showed us that we now have the mosquitoes, our proposed vector, shuttling between humans and possums,' Stinear says. 'A really nice, hard, direct link between the three key elements we were proposing – mosquitoes, humans and possums.'

Keeping the flame alive

The ulcer-hunters also had to prove that the strain of bacteria was the same across all three species involved in this transmission line. PCR testing is very sensitive, but relies on amplifying small segments of DNA. There was a danger that the segments of DNA they were testing were actually from different bacteria strains across people, possums and mozzies. They couldn't grow the bacteria in the lab for sequencing, either.

A new technique called sequence capture enrichment, which utilised an RNA 'hook' to fish out DNA from the mosquitoes, delivered the next eureka moment. 'We used a cutting-edge genomic method that allowed us to pull out, from a single mosquito, almost a complete genome of *Mycobacterium ulcerans*,' Stinear says.

The genomic fingerprint from the bacteria in the possums, mosquitoes and from the ulcers festering on people's feet was a perfect match.

'It really showed that, yes, the bacteria in the mosquitoes is the same strain as the bacteria in humans. It was a "put down the glasses" moment.'

Lastly, the scientists showed there was a non-random geographical association between mosquitoes, possum poo infected with *M. ulcerans* and cases of Buruli ulcers. In Rye, an enclave on the Mornington Peninsula known for cases of the ulcer, all three overlapped. The hierarchy of evidence had its crowning star.

Stinear has gone back to the WHO in Geneva and to other microbiology conferences to present the new research. 'We haven't had any dispute with our discovery,' he says.

Now he and his colleagues are working on new methods of mosquito control to help arrest the rise and spread of Buruli. There's more work to be done – such as investigating why ulcers are spreading faster and wider – but the publication of this latest research, which solves an 80-year mystery, is a satisfying achievement.

'Everyone's kind of kept the flame alive,' Stinear says. 'It's been a really dedicated small team of scientists that aren't willing to give up.'

❋ *In the heart of the forest, one woman built a house of slime*, p. **8**
The heroes of Zero, p. **29**

'WHY WOULD YOU FIND ME ATTRACTIVE?': THE BODY DISORDER THAT NEEDS MORE ATTENTION

Lydia Hales

On a typical day, Eve* would wake up and start prodding at her body while still in bed. She would avoid catching sight of herself while showering, or while trying to find an outfit to best hide her body.

'The day would usually be spent at work thinking about what to eat or not to eat and worrying about how I appeared to others, making sure I was holding in my stomach or standing or sitting behind something. I would check myself in every reflective surface I passed during my entire waking moments: shop windows, bathrooms, cars, mirrors.'

She would then go home alone to cook, which she dreaded, eating as quickly as possible to get it out of the way. She would then spend about an hour looking at her body, pinching and squeezing it.

'I would usually go to bed sad and angry about my appearance, hating it – wishing I would morph into another body and [thinking] my life would be so much easier.'

For Eve, body dysmorphic disorder (BDD) has been a shadow attached to every thought since childhood. She was often praised for being 'pretty' as a child, but the compliments became backhanded as she got older, with suggestions from boys at school, and later men, that she would be more attractive if she were thinner.

'Then I went through puberty and I felt like I changed, like I became a kind of monster.'

Convinced there was something wrong with most of her body – particularly her stomach, thighs and bottom – she embarked on extreme exercise regimes in a bid to change her naturally 'hourglass' shape. At 16, Eve became bulimic.

She was driven by 'shame, self-loathing and wishful thinking – if I can just change this, I can do all the things I want'.

Now in her fifties, Eve says she had a couple of relationships, but the disorder ultimately led her to push people away; over the years she socialised less to avoid the anxiety of being looked at.

'It's a hard one because I really longed for connection ... but if people told me I was beautiful, I'd think there was something wrong with them: "Why would you find me attractive when I'm so revolting?"

'I believed I'd never allow myself to be intimate again as my body was too repulsive, and if a partner saw me naked, they would reject me. I resigned myself to a life of being single.'

What Eve didn't know then was that no amount of exercise, or even surgery, could change what she saw.

Extreme body dissatisfaction

Although BDD is becoming better known, experts say it is often still confused with poor self-esteem or narcissism, sometimes with devastating results. The disorder involves preoccupation with perceived – but nonexistent or minor – flaws, causing enormous distress and varying repetitive behaviours to hide or 'fix' the flaw.

Eighty per cent of sufferers consider suicide and roughly a quarter attempt it.

It is widely cited as affecting about 2 per cent of the general population, making it more prevalent than anorexia and schizophrenia combined. Experts say the true figure could be higher – and rising.

Susan Rossell, an internationally recognised BDD expert, is running a survey to understand its extent in Australia.

'Those prevalence figures are very old and based on [data from] other countries,' says Rossell, a professor of cognitive neuropsychiatry at Swinburne University.

'We've noticed a general trend over the last ten years of an increase in people reporting extreme body image dissatisfaction leading to BDD. But particularly during the pandemic, a range of things there to support us with our everyday lives we know now are quite triggering for people who have a potential to develop BDD.'

An example is 'the Zoom effect'. Rossell and colleagues found one-third of 335 adults surveyed who used video calls during the pandemic reported new concerns about their appearance; they also expressed greater interest in seeking aesthetic procedures.

Dr Katharine Phillips, a professor of psychiatry at New York-Presbyterian and Weill Cornell Medical Center in the US, notes genetics and the environment contribute to BDD's development.

Phillips says the last nationally representative US prevalence survey, conducted in 2015, found it to be higher than in any previous study – at 2.9 per cent.

'But I suspect BDD is becoming more common, perhaps because of widespread use of image-centric social media, which can promote unrealistic beauty standards, enables morphing of one's appearance and can encourage people to compare themselves with very attractive people such as celebrities.'

'We had no treatments'

During her medical residency, Phillips became interested in some severely ill patients – they could not work or socialise and some had attempted suicide because they believed they were so ugly. She had never heard of BDD, nor had her supervisors.

'It had fallen through the cracks of modern-day psychiatry,' Phillips says. 'Even though BDD had been described for more than

100 years, the field knew almost nothing about the disorder ... We had no scales or assessments to diagnose it or assess its severity and we had no treatments. So I started on a quest to learn these things.'

BDD was first mentioned in the widely used Diagnostic and Statistical Manual of Mental Disorders in 1980. Phillips, whose research influenced diagnostic criteria and descriptions in several later editions, says recognition has improved compared with when she began research in the 1990s, but more is needed.

'BDD has been much less studied than many other severe psychiatric disorders and we need more research studies on virtually every aspect of the disorder. We especially need data from diverse populations and in children and adolescents.

'It tends to be considered less important than other severe mental disorders, even though it's more highly associated with suicidal ideation and suicide attempts than most other serious mental illnesses.'

It often occurs alongside other mental health issues, but many sufferers avoid discussing their symptoms out of shame.

Perceived ugliness

Studies by Rossell and colleagues including David Castle, a psychiatrist and professor of psychiatry at the University of Tasmania, show changes in brain structure and connectivity among BDD patients.

Castle says while obsessive-compulsive disorder (OCD) and BDD both involve obsessional thinking and ritualised behaviours, BDD is distinct in important ways.

'Most people with OCD recognise their concerns are excessive, but in BDD, about half ... have delusional convictions about their perceived ugliness: they completely believe it.'

There are also differences in visual processing: BDD sufferers tend to misinterpret facial expressions, being more likely to rate the expressions of onlookers as contemptuous or angry.

Rossell says unlike the brains of healthy people, which switch between looking at fine detail and at the big picture, the brains of people with BDD get stuck in detail mode. If anyone stares at one part of their appearance for a long time, their perception of it will become distorted, and this is thought to be at play in BDD.

Cosmetic quests

The Melbourne-based clinical psychologist Dr Toni Pikoos says about three-quarters of BDD sufferers seek cosmetic or dermatology procedures. But, she says, these are unlikely to help and often present increased risks to patients and providers.

'They can be vulnerable consumers. Often they're quite desperate, so they're willing to pay the money to feel better ... Unfortunately, in up to 90 per cent of cases of people with BDD, they don't experience any change in their symptoms afterwards.'

Pikoos often sees clients who were initially satisfied with a cosmetic surgery, but months or years later feel the 'flaw' has come back or become worse.

'They can become quite depressed and fixated on that area of their appearance, which can sometimes lead to getting really angry at the surgeon ... There's a [US] statistic that 29 per cent of people with BDD have complained or taken out litigation against their cosmetic practitioner.'

Following an inquiry into the cosmetic industry, in July 2023 Australia's medical regulator, AHPRA, tightened guidelines for doctors providing cosmetic procedures, with further changes expected for providers of non-surgical procedures – including nurses, dentists and Chinese medicine practitioners.

The changes require doctors performing cosmetic surgery or prescribing cosmetic injectables (such as the botulinum toxin and dermal fillers) to screen patients for BDD and refer at-risk patients for external assessment.

Pikoos, who co-founded an organisation to help cosmetic

practitioners with psychological assessments and consulted on the AHPRA review, says there has been an uptick in referrals for assessment since July, some of which have needed further treatment or support.

Access to treatment

BDD symptoms are unlikely to improve without targeted treatment, Castle says, usually involving high doses of SSRI antidepressants combined with cognitive behaviour therapy focusing on avoidance and safety behaviours, including work with mirrors.

'You can change people's lives completely. The tragedy is, often people are coming to us very late and having gone through a lot of cosmetic procedures.'

Rossell estimates there are a dozen mental health professionals in Australia trained to treat BDD, many of whom she trained, but laments that there is no specialist focus.

BDD is not a searchable issue in the Royal Australian and New Zealand Royal College of Psychiatrists (RANZCP) or the Australian Psychological Society (APS) clinician directories, although APS includes body image as a subcategory of 'personal' issues. Rossell says improving searchability is 'crucial'.

She is concerned patients – and those who want to refer them – struggle to find the existing specialists and she urges any practitioners with experience in BDD to notify RANZCP and APS, since they rely on information provided by practitioners.

The RANZCP president, Dr Elizabeth Moore, says research shows the disorder is poorly understood and that prevalence could be higher due to underdiagnosis. 'Amongst health professionals, there are concerns around missing symptoms associated with BDD in patients, or misdiagnosing BDD patients for depression, social anxiety or OCD.'

Eve* estimates she saw ten psychologists or psychiatrists before being diagnosed with BDD three years ago. She has been receiving specialist treatment for almost a year, which has been transformative.

'I wake up feeling positive and the focus on my physical appearance has become less intense.'

It is also bittersweet.

'It's almost surreal to have someone who understands it. It's validating, but it also made me a little angry that it had taken so long to get help and that I've spent most of my life grappling with it myself – the shame and stigma and being dismissed.'

However, she believes this progress is the reason she is now in a supportive relationship – marking the first time she has told a romantic partner about her BDD.

'Things have changed drastically ... and I am in love. I have opened myself up to the most caring and understanding person I have ever met.'

Eve is now studying psychology, with plans to specialise in the treatment of BDD.

*Name has been changed.

✱   *How scientists solved the 80-year-old mystery of a flesh-eating ulcer*, p. **145**
Why my adenomyosis went undetected for five years, p. **157**

WHY MY ADENOMYOSIS WENT UNDETECTED FOR FIVE YEARS

Lauren Chaplin

It's a debilitating condition that leaves a third of Aussie women in frequent pain. So why have so few people heard of it?

'I finally have an answer for you,' my doctor said as I settled in a chair in her tiny office. I almost didn't believe her.

It had been five years since I first sought advice for the crippling periods, pelvic pain, bloating, fatigue and nausea that had left my body perpetually feeling like a beat-up punching bag.

Until now, I had been dismissed by multiple doctors and specialists. One told me it 'might be PCOS' (it wasn't), others said it was 'just bad period pain'.

I'd read stories of other women experiencing the same types of symptoms and not being able to find a diagnosis. So when the doctor said she had an answer, I felt the relief course through my body as I realised my experience was finally about to be validated.

'Your ultrasound showed that you have adenomyosis,' my doctor said.

I stared at her blankly.

'What the hell is that?'

A painful journey to a diagnosis

The first time I had an adenomyosis flare up, the pain was so bad my partner thought my appendix had burst. Even his attempt to carry me to the car left me in complete agony, forcing me into a ball and leaving me nauseous. We settled for calling a doctor out to our apartment. We were promised a visit within two hours, but no one came.

The next morning, I tried to explain what I'd felt to my local GP. He told me it was 'probably just a urinary tract infection' and sent me away with a script for antibiotics. Knowing I had none of the symptoms of a UTI, I went to a different doctor, who ruled it out almost immediately.

This GP sent me for a pelvic ultrasound but, when the results were inconclusive, simply said: 'I'm sorry, I'm not sure what else I can do for you'.

This trend of disbelief and medical gaslighting followed through three more GP visits, countless blood tests, another ultrasound and, finally, a gynaecologist visit that left me in tears.

That appointment began with the insinuation I was overdramatising my pain, quickly followed by an accusation that I wanted 'surgery I didn't need' when I requested a laparoscopy to rule out endometriosis. When I picked up my bag to leave, the parting shot was delivered: 'Some women just have pain. We don't know why. You might have to learn to live with it.'

I lived in pain for three more years, before another GP grew suspicious of my symptoms and sent me for another ultrasound. I wasn't feeling hopeful, but it finally gave me the answer I needed. Unfortunately, the diagnosis was just the beginning of a long journey to manage the pain.

What is adenomyosis?

Often referred to as the 'evil cousin' or 'evil sister' of endometriosis, adenomyosis is a disease that manifests in a similar way, with cells like those that line the uterus growing where they aren't supposed to.

In endometriosis, they can be found anywhere outside of the uterus, which can cause severe pain and infertility. In adenomyosis, the cells and their supporting tissue, called stroma, are found within the muscular wall of the uterus.

It's estimated that around 35 per cent of women have adenomyosis, yet surprisingly little is known about the disease.

While some women don't experience symptoms, many others report painful periods, heavy or abnormal bleeding, pelvic pain and painful sex. But it doesn't end there.

'There's more and more evidence to support adenomyosis' impact on infertility, and possibly also obstetric outcomes for increased risk of miscarriages,' Dr Samantha Mooney, an obstetrician, gynaecologist and clinical researcher, told news.com.au.

'It can also have an impact on both spontaneous fertility and IVF success rate. Up to a quarter of people who suffer infertility will have a diagnosis of adenomyosis,' she said.

It's also extremely common for adenomyosis and endometriosis to go hand in hand, with some studies suggesting up to 89 per cent of people with endometriosis also have adenomyosis.

Because women have long had their experiences dismissed, both endometriosis and adenomyosis typically take years to diagnose – such as in my case – which can result in symptoms getting worse.

Excruciating delays

Perhaps the biggest reason women are struggling to be formally diagnosed with adenomyosis is that specialists can't form a unanimous decision on exactly what it is.

It was originally seen as a disease that affects those over the age of 40 after childbirth, and hysterectomies used to be the gold standard in a diagnosis for adenomyosis. As more and more women in their twenties and thirties have been diagnosed, this has become less viable.

Instead, imaging techniques such as magnetic resonance imaging (MRI) and transvaginal ultrasonography (TVUS) are most commonly used to find the disease. However, this isn't always straightforward either.

'We can't find agreement yet in the international literature as to exactly what constitutes adenomyosis on ultrasound,' Dr Mooney said. 'There's a group that published something, think of it as a consensus on how we should report an ultrasound, to say whether or not certain features are there with regards to adenomyosis. But even findings from these often very high-tech and skilled ultrasounds, we can't agree whether they're absolutely specific to adenomyosis or not.'

It's the same for MRIs.

No relief

Even once you have a diagnosis, things don't necessarily get any easier.

It took me multiple attempts to find a pain medication that offered me even the slightest bit of relief. Some days, it's still not enough.

Then there's the first-line therapy for managing symptoms. Like most conditions that solely affect women, it comes in the form of the contraceptive pill. Once again, it can take a lot of trial and error to find one that works. For those looking to get pregnant, it's not even an option.

As Dr Mooney points out, there's also an ongoing cost to it not everyone can afford.

'Only very few of the hormonal remedies that may treat adenomyosis, or at least keep the symptoms at bay ... are covered by the Pharmaceutical Benefits Scheme (PBS),' she said. 'Adenomyosis isn't even listed on the PBS coverage for some of the medications that we use.'

Some surgical procedures may also be considered for adenomyosis, with a hysterectomy used as a last resort for those who haven't responded to any other treatment.

A painful conclusion

There's a relief to finally putting a name to your pain.

Unfortunately, this relief is snuffed out moments later with the words 'there's currently no cure for adenomyosis'.

Some attempts are being made to improve the diagnostic process, such as a letter submitted this week to the *Journal of Clinical Pathology* calling for pathologists to form a consensus on exactly how to diagnose the disease.

But if real change is going to happen, we need to start talking about it.

Endometriosis is finally starting to get the recognition it sorely deserves.

It's about bloody time adenomyosis did too.

✱   *'Why would you find me attractive?': The body disorder that needs more attention*, p. **150**
'Give the espresso a little swirl': The very particular science of a good cup of coffee, p. **250**

DOING DRUGS DIFFERENTLY: FOR PUBLIC HEALTH, NOT PROFIT

Clare Watson

When pharmaceutical giant Johnson & Johnson got the tick of approval for its drug bedaquiline on the last day of 2012, a new chapter of tuberculosis treatment began. Bedaquiline was the first antibiotic developed for the bacterial disease in nearly 50 years – a breakthrough millions of people across Asia and Africa had been crying out for.

A decade later, bedaquiline has become a latest flashpoint in the searing debate about affordable access to lifesaving medicines. Knowing its main patent for the drug would expire in July 2023, Johnson & Johnson sought to extend its monopoly by enforcing secondary patents for an ever so slightly altered version in more than 65 countries.

India took a stand. Its Patent Office rejected the company's application in March 2023, following a four-year global campaign agitating for fair access to the drug. Then, in July 2023, a historic deal permitted nonprofit organisation Stop TB Partnership to supply cheaper generic versions of bedaquiline to 44 low- and middle-income countries.

While some countries with the highest burden of tuberculosis aren't covered by the agreement, those that are will be able to buy the drug at half price. The Stop TB Partnership estimates that by the end of 2024, more than 51 000 extra treatments could be purchased with the cost savings.

The deal is a small concession from a pharmaceutical giant that has reaped the benefits of market exclusivity for a decade, and experts say it's long overdue. 'It raises the question, why couldn't this have been done earlier?' asks Diego Silva, a bioethicist at the University of Sydney concerned with infectious diseases. 'We didn't need to get to the eleventh hour of a patent expiring for this outcry to happen.'

The story of bedaquiline is just one example of an industry that has long prioritised profit over public health. Companies argue that sky high prices are necessary to fund research and development (R&D) into new drugs, but the data doesn't back up this claim. Plus, researchers have repeatedly exposed the pharmaceutical industry's profit-driven motives that shape its clinical trials, skew its drug development and inflate drug prices.

But does it have to be this way? What – or where – are the alternatives?

Around the globe, a few initiatives are going against the grain of the for-profit pharmaceutical industry, instead prioritising neglected diseases, refusing to patent their drug discoveries and funding research that puts public health first. Their work to develop new medicines for a fraction of the cost shows that other models of drug development are not only possible – they're succeeding.

So how do they work, and can we bring these ideas home to Australia?

Fact-checking R&D spending

In a global industry that spent US$141 billion on R&D in 2015, it's hard to get a clear picture of what it actually costs to develop a drug. 'Drugs are expensive to produce, but part of the problem is we're not entirely sure just how expensive it is,' Silva says.

The industry isn't known for its transparency. Companies protect their research products with a thick web of patents, and in financial reports they tend to average out their R&D spending across the drug development pipeline.

Researchers like Joel Lexchin, a pharmaceutical policy researcher at York University in Toronto, Canada, analyse the practices of drug companies based on what data is publicly available. It's enough to see trends emerge. 'R&D gets sacrificed to share prices,' Lexchin says bluntly.

While companies do bring to market some truly innovative medicines – like bedaquiline – their business model has changed. Where once they would reinvest profits into R&D, Lexchin says the world's largest companies now spend more money on marketing their products, buying back their own stocks to lift share prices and paying shareholder dividends.

Health economist Aris Angelis and colleagues laid out the costs in the *British Medical Journal* in early 2023. Based on financial reports from 1999 to 2018, the 15 largest pharmaceutical companies spent nearly twice as much on 'selling, general and administrative activities' as they did on R&D: US$2.2 trillion compared to US$1.4 trillion over 20 years. Most of those same companies spent US$577 billion on share buybacks and dividends from 2016 to 2020 – US$56 billion more than R&D during that time, according to a 2021 US government drug pricing investigation.

Rather than investing in discovery research, large companies have also taken to buying up small start-ups that have done the hard yards developing new candidates. For instance, US biopharmaceutical company Gilead Sciences didn't discover sofosbuvir – an antiviral treatment that transformed hepatitis C care a decade ago. It bought the drug from a start-up for US$11.2 billion. According to a US Senate inquiry, Gilead recouped nine-tenths of that amount in its first year of selling the drug. 'That's the model drug companies are using these days,' Lexchin says.

The industry claims to be innovating new medicines, which are priced to recoup research costs and offset other failures. R&D budgets of the 14 leading pharmaceutical companies grew in the two decades to 2018. But a 2021 systematic review of 19 studies found that companies are spending more for each new drug they produce; in other words, pharmaceutical R&D has become more inefficient over time. Another peer-reviewed study traced this decline back to the 1950s.

Research shows that clinical trials are getting longer and more complex, and failure rates in drug development are rising as the industry pursues high-risk, high-reward medicines. Just one in ten drugs that entered early stage clinical trials between 2003 and 2011 got approved.

Drug developers also can't learn from their competitors' mistakes, because their work is shielded by patents and companies tend to suppress negative results, so each sinks money into well-trodden paths that lead to repeated failures. For instance, a 2023 study in *JAMA Network Open* found that starting in the early 2000s, drug companies spent US$1.6 billion to $2.3 billion on 183 cancer trials – involving more than 12 000 patients – to test 16 drug candidates for a popular cancer target. None were approved for treating cancer.

Pharmaceutical companies are also quick to abandon a drug candidate if business priorities change, Lexchin says. The result: companies are outlaying more money than ever before for fewer new drugs – so when they do hit on a breakthrough, patients are paying the price.

The most lucrative medicine on the market, prior to the Covid-19 pandemic, was a treatment for rheumatoid arthritis called adalimumab. Its manufacturer, AbbVie, spent an estimated $US14.7 billion on R&D, then grossed ten times as much globally after the drug's approval in 2002. Researchers estimate that this income translates to the company netting an eye-watering US$110 billion in excess profits – over and above what would be considered a fair margin for the amount it spent on developing, producing and marketing the drug.

Those kinds of blockbuster drugs are a costly exception. According to economists at the Tufts Center for the Study of Drug

Development in Boston, it costs US$2.6 billion to bring one new medicine to market. This includes lab research and clinical trials, but also accounts for other drugs that don't make it through to approval and other financial losses.

But leading pharmaceutical policy researchers have labelled that US$2.6 billion figure an inflated estimate, pointing out that it's based on confidential data on just 106 drugs from ten pharmaceutical companies that no one can independently verify.

Subtract the tax breaks companies received and account for the fact that only the costliest 20 per cent of drugs were included in the analysis, and researchers have revised the number to about one-tenth of the industry's claimed total. Other more recent analyses of publicly available data have found that on average, the development of one drug costs less than one-third to about half as much as the Tufts estimate.

What's worse is that the majority of newly approved medicines hardly improve on existing drugs, if at all, says Barbara Mintzes, a pharmaceutical policy researcher at the University of Sydney. Mintzes calls them 'me-too' drugs: new formulations of old compounds that provide little advantage over current drugs but which serve to prolong patent protection.

Patents are designed to reward drug developers for innovating new products, giving them exclusive rights for a set period of usually 20 years. However, few new drugs are truly innovative. More than half of new drugs are no better than existing treatment options, according to yearly investigations from French organisation Prescrire International. Another 15 per cent of approved drugs are actually worse: either less effective or poorer safety-wise, says Mintzes.

'When many new drugs do not offer any therapeutic gain to patients, the only beneficiaries are the companies that are marketing them,' Lexchin remarked in a recent commentary in the *Journal of the American Medical Association*.

So if pharmaceutical companies aren't as innovative as we've been led to believe, and if the innovation they do provide comes at such a high cost, is there another way?

Exploring the alternatives

Since its inception in 2003, the aptly named Drugs for Neglected Diseases Initiative (DNDi) has developed a dozen treatments for six deadly diseases – with no labs and for a fraction of the industry cost.

DNDi was founded by medical humanitarian organisation Médecins Sans Frontières (MSF) in collaboration with research institutes in India, Brazil, Kenya, Malaysia and France, after MSF realised it often didn't have the medicines it needed to save lives.

The initiative focuses on advances in patient care for neglected tropical diseases such as dengue fever, malaria and leishmaniasis. These collectively affect nearly 2 billion people worldwide, yet represent only 0.5 per cent of the more than 56 000 candidate products currently in commercial development.

DNDi operates like a virtual biotechnology company. It contracts industry collaborators and academic partners to conduct specific studies at each stage of its drug development pipeline, funded by in-kind donations and philanthropy. For 12 years, Australian-born medicinal chemist Robert Don was at the helm of DNDi's drug discovery pipeline. Patients were front of mind in every research phase, and 'part of every decision we made', Don says.

One of Don's proudest achievements sums up the initiative's mission. DNDi developed a drug for African trypanosomiasis (also known as sleeping sickness), an often fatal parasitic disease. The medication isn't registered yet but may one day replace the existing treatment, melarsoprol, which has to be injected over weeks to months. 'Patients would flee the clinic because it was so painful,' recalls Don. 'We finally got that down to a single pill that had the same side effects as an aspirin.'

After two decades in operation, DNDi estimates it spends US$4 million to $34 million to develop and register treatments that combine or repurpose existing drugs. Developing an entirely new chemical entity costs US$63 million to $200 million. At its most expensive, that's still 13 times less than the industry estimate.

Those new chemical entities come from trawling through the huge libraries of chemical compounds that pharmaceutical firms amass. 'It took us years [of negotiations] to break in with the first company,' Don says.

But once DNDi was granted access, its scientists could screen thousands of chemical entities using robotic assays to see if any were effective in killing pathogens grown in lab culture dishes. They only pursued a promising drug lead if it could be made as tablets, which are easier and cheaper to distribute in remote, humid regions.

Given the diseases it targets, DNDi runs clinical trials in some extremely challenging environments, crossing rivers and rainforests to reach remote clinics in Africa, Asia and the Americas. But their trials are made somewhat easier by the fact they generally aren't looking for incremental improvements between lookalike drugs as pharmaceutical companies do. DNDi seeks clear improvements in patient care, which are more evident in smaller trials, and smaller trials with fewer patients help to keep costs down.

DNDi also rarely patents its discoveries. To ensure equitable, affordable access to its medicines, the initiative stipulates in its negotiations with drug companies and research partners that their products must be free of any restrictive patents and sold at minimal cost, in all endemic countries, regardless of income levels.

Sharing is caring

DNDi stands in stark contrast to the pharmaceutical industry, but it's not the only alternative model. In Europe, Italy's Mario Negri Institute also prides itself on making its research accessible.

Founded in 1963, the Institute was the idea of Italian pharmacology researcher Silvio Garattini, whose working-class background led him to envision a medical research institute devoted to the public interest. He sold local philanthropist Mario Negri on the idea, and it came to life.

Based in Milan, the Institute aims to improve health with independent, transparent science. It never patents its discoveries, and all its findings are publicly available – including failures. It eschews placebo-controlled trials, instead designing trials to test if new therapies improve on existing treatments. It also investigates harmful side effects that might otherwise go unreported.

Just like any other research organisation, Mario Negri pieces together government grants, industry funding and public donations to fund its work. However, it maintains staunch independence from the pharmaceutical industry and governments by ensuring no funding source amounts to more than 10 per cent of its annual budget. This allows it to pursue research, design trials, analyse data and share its findings freely.

'Open sharing of science can lead to advances for all of us much more quickly. We certainly saw that during the pandemic,' Mintzes says. 'The Mario Negri Institute is an example that that kind of model can exist and actually flourish.'

Where DNDi focuses on select neglected diseases, Mario Negri has a wide-ranging program that tackles some of the biggest health problems of our time, including cardiovascular disease, cancer and neurodegenerative diseases.

The Institute also holds the manufacturing industry and governments to account through investigations of environmental pollution and contamination.

It's also beating the pharmaceutical industry at its main game: large-scale clinical trials. In the 1980s, Mario Negri ran some of the first 'mega-trials' in medicine, which revolutionised clinical trial design. The first of those trials showed an inexpensive treatment administered quickly could prevent deaths from heart attacks. The trial involved nearly 12 000 patients across the Italian healthcare system, yet it was planned, conducted and published in under three years. These days, Mario Negri can run trials at one-tenth of the cost per patient of standard industry trials.

Subscription medicine

We aren't short of options to change the way we approach drug development. There are plenty of other levers that governments could pull to reorient clinical research towards the areas of greatest need, to prioritise public health over profits.

India is a prime example: its interpretation of patent laws has enabled the country to reject patents from pharmaceutical companies for drugs that do little to improve on existing therapies on multiple occasions.

Before it joined the European Union, Norway also had similar laws to ensure approved drugs were either more effective, easier to take or had fewer side effects than available treatments.

Lexchin says Australia could likewise change its patent laws and tighten up drug regulations to only permit drugs 'that really make a difference'. However, the pharmaceutical industry wields strong influence over many national governments and drug regulators. The industry sustains huge parts of national economies in countries like the US, and largely funds regulatory agencies through user fees. As Mintzes says, it would take 'quite a bit of bravery and innovative thinking to bring in these kinds of policies'.

Although Australia only represents a tiny slice of global drug spending and is too small to influence the research interests of the pharmaceutical industry at large, we're a wealthy country and could steer R&D by channelling more public funding into specific areas of national need.

'Pharmaceutical companies don't exist without research that is publicly funded,' the University of Sydney's Silva points out.

'R&D isn't just the moment when a compound enters phase I testing.' It starts long before that, in university labs and research institutes. Reorienting the R&D pipeline begins with adequate funding for basic science, and in Australia research funding has stagnated over the past decade, reducing the likelihood of chancing upon new drug candidates.

Lexchin agrees that increasing public funding, especially of clinical trials but also of early-stage research, could yield better outcomes. 'The public sector plays a much larger role than is currently recognised,' he says.

In fact, 25 per cent of new drugs originate in the public sector – and those drugs have more therapeutic value than the ones coming from industry. Take bedaquiline: researchers estimate that the public sector invested US$455 million to $747 million in the drug's development – three to five times as much as Johnson & Johnson spent.

'In many cases, the public will pay twice,' says Mintzes: once for the initial, public investment in a drug's development, and again when governments subsidise its access because the prices set by drug companies are so high. 'We should be incensed,' adds Silva.

Some policy experts argue drug prices should be capped or early access guaranteed if those medicines were developed with large amounts of public funding. High prices can restrict access to medicines in the US or Australia, as much as any other country. For example, when Gilead priced their hepatitis C antiviral sofosbuvir at US$84 000 for a course of therapy – a blistering US$1000 per pill – less than 3 per cent of eligible Americans could access the treatment through Medicaid.

Australia actually had a radical answer to that problem: in 2015, it pioneered a lump-sum payment fee to manufacturers – A$1 billion over five years – in exchange for an unlimited supply of seven hepatitis C antivirals, including sofosbuvir. Researchers estimate that the government saved A$6.42 billion in those five years and treated 93 413 more patients than if they had paid per packet.

This unconventional approach, dubbed the subscription or Netflix model, has since been adopted by the UK to spur innovation in antibiotic R&D, an area of development that has slowed to a trickle, and Sweden is trying out its own subscription program. Time will tell how effective these pilots will be.

Patent swaps are another idea. Pharmaceutical companies would forgo patenting an essential medicine needed in low- and middle-income countries in exchange for a patent extension on a non-essential product sold elsewhere.

It seems unlikely that this piecemeal approach – one patent here, one drug there – will reshape the global R&D landscape, but each of these strategies offers gains in areas of huge need. Progress can come from many small steps, as well as giant leaps.

Bringing it home

If there's one area where Australia could really take the lead, it would be in the fight against Group A streptococcus bacterial infections. While we've made great strides in quashing dengue fever and reducing tuberculosis, group A strep is a different story. It's among the world's deadliest pathogens, causing a whole spectrum of illness and disease, from sore throats to flesh-eating necrotising fasciitis. Yet still there is no vaccine.

The Australian Strep A Vaccine Initiative (ASAVI) is hoping to change that. The initiative formed in 2019 with an A$35 million windfall from the Medical Research Future Fund and the clear goal of progressing a strep A vaccine to phase II clinical trials in the next five years.

It's an example of another R&D model gathering speed: mission-oriented initiatives that 'work on a very specific research problem over a defined period of time, deliver results and then move on,' explains Daniel MacArthur, a population geneticist at the Garvan Institute of Medical Research in New South Wales.

Repeated or untreated strep A infections can permanently

damage the heart, a condition called rheumatic heart disease, which leads to heart failure and stroke.

Australia has one of the highest rates of rheumatic heart disease in the world, particularly in one part of our population. 'There is such an enormous burden of rheumatic heart disease, particularly in our First Nations people,' says immunologist and ASAVI project lead Alma Fulurija.

Aboriginal and Torres Strait Islander people account for more than 90 per cent of cases of rheumatic heart disease, and are nearly 20 times more likely to die from the condition than the general population.

Fulurija says ASAVI was created as a 'new way of accelerating vaccine development' in an area 'that perhaps industry wasn't as interested in'.

Fewer than 12 vaccines are in early development for group A strep, compared to hundreds in the pipeline for HIV, tuberculosis and malaria, she says. (That's partly because human trials into strep A vaccines were prohibited for nearly 30 years after the US drug regulator got spooked by safety data from an early study. The ban was lifted in 2006, but, still, a void of industry investment remains.)

According to Michael Good, a vaccine researcher at Griffith University, no one is interested, commercially, in making a vaccine to prevent rheumatic heart disease because it mostly affects lower-income countries, which are not the most profitable markets. Pharmaceutical companies may be interested in a vaccine for tonsillitis or strep throat that could be sold in wealthy countries too, Good says, 'but that's not the main reason we're in this game'.

Penicillin and other antibiotics remain effective against strep A, although some strains are developing resistance. Since the 1990s, Good has been trying to develop a vaccine to protect against strep A infections – and thereby rheumatic heart disease – by scratching together grants and philanthropic funding to sustain his group's research. Roughly A\$20 million and three decades later, they have a vaccine candidate in a phase I safety trial of 45 volunteers.

It's at this point that university-led research so often stalls – and why ASAVI could be critical. 'When you step out of discovery research and move into development, it's a different kettle of fish,' says Fulurija, who spent 15 years working in the pharmaceutical sector. 'What ASAVI can do is bridge those two.'

Another world is on its way

Organisations like DNDi and the Mario Negri Institute demonstrate that it is possible and beneficial for drug research and development to prioritise public health over profit. But these initiatives and other alternate programs didn't spring from thin air – changing an industry takes concerted effort on many levels.

'There needs to be pressure from below, from the public, from clinicians and from researchers – and there needs to be political courage from above to make changes,' Lexchin says.

And sometimes, this pressure gets real results. In late September 2023, two months after Johnson & Johnson announced its historic deal, the company dropped its patents for bedaquiline. The company will no longer enforce its secondary patents for the tuberculosis drug in 134 low- and middle-income countries, which represent 99 per cent of global tuberculosis cases. Manufacturers can now make and supply generic versions of bedaquiline years before the secondary patents expire in 2027.

Five years ago, the drug cost in-need countries US$67 per patient per month.

Competition between generic manufacturers is expected to bring prices down to US$8.

✱ *The heroes of Zero* p. **29**
   Science in the balance, p. **208**

INDIGENOUS SCIENCE MUST BE A STANDALONE NATIONAL SCIENCE POLICY

Joseph Brookes

Indigenous knowledge has been omitted as a standalone priority from a draft version of Australia's new national science and research priorities, creating a 'massive gap' that experts say will limit funding for an already under-resourced and underutilised area.

Indigenous researchers and a peak group for scientists are now calling for a late change to the year-long update process that would make the elevation and investment in First Nations perspectives on science, technology and innovation a standalone priority within the new science strategy.

Kamilaroi water scientist Associate Professor Brad Moggridge said the stated integration of Indigenous knowledge throughout the new draft priorities is welcome. But its absence as a standalone priority leaves a 'massive gap' that will make progress harder.

'Science has had a long time to do things better and I don't think they've succeeded yet, especially in Australia's short history of post-colonisation of 235 years,' he told InnovationAus.com. 'It hasn't done well, and some of the decisions that are being made are poorly informed and I think Indigenous knowledge can value-add and strengthen decisions, but also be a part of the solution to fix 235 years of abuse of country.'

The national science and research priorities are intended to focus the Australian government's support for science and research on the most pressing issues facing the country. The priorities have not been updated since 2015.

'It's really important to understand that these priorities will crystallise what's important for the nation for the decade ahead and guide Australia's research agenda and focus in that coming decade,' said Science and Technology Australia chief executive Misha Schubert, whose organisation represents 110 000 scientists and technologists.

'[The priorities] will also play a key role in determining what research gets funded, which is why this is such a crucial opportunity and why we must not miss this moment in history', she said.

Industry and Science minister Ed Husic last year launched a review of Australia's 'out-of-date' national science and research priorities, which are currently: food, soil and water, transport, cybersecurity, energy, resources, advanced manufacturing, environmental change and health.

Despite flagging the elevation and investment in First Nations perspectives on science, technology and innovation from the start, draft priorities released last month do not have this or similar as a standalone priority.

'Aboriginal and Torres Strait Islander peoples asked that First Nations knowledge and knowledge systems be reflected throughout the priorities, rather than as a standalone area,' according to Australia's draft National Science and Research Priorities report.

However, others involved in the draft priorities have also told InnovationAus.com the omission of a standalone priority was a surprise, given it was at the forefront of early discussions.

Asked about the change, a spokesperson for Mr Husic said the government is 'committed to elevating First Nations science'.

'We are currently considering submissions provided in response to the draft priorities from a range of stakeholders,' the spokesperson told InnovationAus.com. 'We will have more to say once this process is concluded.'

Science and Technology Australia's call to reinstate it as a standalone priority is backed by Professor Moggridge, who also chairs the National Indigenous STEM Professionals Network, and other Indigenous science leaders like CSIRO Board Director Professor Alex Brown and founder of Deadly Science Associate Professor Corey Tutt.

Other science and research groups have not been as critical but many have called for more detail on how Indigenous knowledge will be recognised and interwoven throughout all priorities.

The government report flags further discussion with First Nations communities to ensure 'their expertise is integrated respectfully and where appropriate'. It sought more input in the final round of consultation that closed last month on four draft priorities:

- ensuring a net zero future and protecting Australia's biodiversity
- supporting healthy and thriving communities
- enabling a productive and innovative economy
- building a stronger, more resilient nation.

Indigenous knowledge is only explicitly identified as an aim and research area in the draft report for the net zero and biodiversity priority, but the government is seeking more feedback on building a system that recognises and values First Nations knowledge and knowledge systems.

'Looking deeper into the draft priorities,' Professor Moggridge says, 'it has Indigenous people only as helping with flora and fauna, and restoring of biodiversity. We're more than that.'

He said more fundamental change is needed to recognise Indigenous knowledge stretches beyond conservation and bio-diversity, but also to ensure it is actually supported and utilised.

'Science needs to evolve and see Indigenous knowledge as an equal, not as myths and legends and stories. I'm not inspired by these priorities,' he said.

University of Melbourne Deputy Vice-Chancellor (Indigenous) Professor Barry Judd, who leads the institution's Indigenous-focused aspirations, is also backing a standalone priority that represents Indigenous knowledge.

'Our research and science system is just beginning to recognise the potential of Indigenous knowledge,' he told InnovationAus. com. 'There needs to be a strong, national focus on Indigenous knowledge, along with dedicated resources to build the Indigenous researcher pipeline, empower Indigenous communities to become equal research partners and enable institutions to enhance the competencies of all staff in all aspects of Indigenous knowledge and knowledge systems.

'While we have witnessed change in the last decade regarding the valuing of Indigenous knowledge, identifying Indigenous knowledge as a standalone priority is an important step towards full recognition of Indigenous knowledge, expertise and relationships.'

Ms Schubert said it is critical to have Indigenous knowledge recognised as a dedicated priority because when operationalised, the priorities act as a reductive list that directs major grant funding.

It would send a 'powerful signal' to funding bodies that have typically overlooked or underfunded the area that Indigenous research and knowledge is fundamentally important to the nation, she said.

'It's what makes our knowledge, our science, our research efforts genuinely unique in the world. So if we missed that moment, we will have missed that chance to embed a powerful signal back into the funding research architecture of the country.'

Deputy Vice-Chancellor Judd, also a Professor of Indigenous Studies, agreed the recognition would help stimulate funding while sending an important message to prospective students and researchers. 'It sends an important signal to Indigenous communities of the value of Indigenous knowledge, leading to the creation of reciprocal collaboration, partnerships and knowledge exchange,' he said.

Professor Moggridge said the current lack of recognition is contributing to an underutilisation of Indigenous knowledge in Australian science, which is reflected in his area of expertise, water.

'Talking as a as an Aboriginal scientist, I struggle to see my relevance and I just keep pumping out the research. It's a very niche space that I'm in, but when you think about water – we are the driest inhabited continent on earth and Aboriginal people don't have a say in how water is managed.'

Professor Judd said Indigenous knowledge has not been properly harnessed and still holds potential for addressing challenges that range from climate change and biodiversity loss to public health and ethical AI.

'There remains huge opportunities to strengthen the relationship between western science and Indigenous knowledge systems and address a multitude of local and global problems,' he said. 'A major barrier is the Indigenous researcher pipeline. If the government is serious about unlocking the Indigenous knowledge that will deliver innovation and economic benefits, they must commit appropriate funding and incentives.

'Priorities must be accompanied by a commitment to building Indigenous research capabilities in key areas.'

Professor Moggridge also sees untapped potential because of the current system and its funding constraints.

'There's Indigenous scientists out there doing the hard yards, but there's also knowledge holders that aren't recognised for the connection and understanding of country,' Professor Moggridge said.

✱ *Doing drugs differently: For public health, not profit*, p. **162**
Indigemoji, p. **271**

ARE PSYCHEDELICS A TREATMENT FOR LONG COVID? RESEARCHERS PROBING THIS MYSTERY DON'T HAVE THE ANSWER YET

Rich Haridy

It was March 2020 and Ash was the healthiest she'd been in 15 years. She had just started an exciting new job and Covid was still a nameless 'novel coronavirus' mainly appearing on cruise ships. One evening, after getting home from the gym, Ash was suddenly struck with a wave of feverish delirium. She passed out and eventually came to a couple of hours later on the kitchen floor with her dog staring down at her.

The next two weeks were a blur, but eventually Ash started to feel better. About a month after the initial illness she had pretty much recovered. And then things started getting strange. She had this feeling her teeth were rotting. A painful pressure began building in her head.

'And it just took over my nerves,' Ash explained in a conversation with *Salon*. 'About six weeks after Covid, I started losing the use of my hands.'

Everything from opening a ziploc bag to using scissors became profoundly difficult. Multiple GPs, dentists, clinical specialists and even a Chinese acupuncturist all had no idea what was going on. By the end of 2020 Ash had stopped working altogether. Alongside the neuropathic problems, all the now common neurological long

Covid issues had become entrenched: brain fog, dissociation, extreme fatigue, memory troubles.

Alienated by mainstream medicine's denial of her condition, Ash became her own guinea pig for the next couple of years. With a deep knowledge of science and a pool of friends in the entheogenic community, Ash tried anything and everything to overcome her debilitating symptoms. Steroids, low-dose naltrexone, melatonin, lecithin, goldenseal, sceletium and a whole world of anti-inflammatory botanical ferments like kefir. Some helped temporarily, some didn't help at all. Ash kept a detailed treatment diary, tracking the effects of everything she consumed.

'People were just sending me random obscure stuff. And I'm like, yep, that doesn't work. That works. That doesn't work. Oh, that doesn't work for more than three days.'

Then in early 2023, Ash tried something completely different. Something she described as a game changer for her condition: a powerful hallucinogenic plant called iboga that originates from Africa. Its active ingredient is known as ibogaine and it's being explored for addiction treatment. It's not clear yet if it will really help – but even more questions remain about its potential for alleviating long Covid.

The serotonin hypothesis

It has been estimated that at least 65 million people have, or have had, long Covid. The illness encompasses hundreds of different symptoms and researchers are still struggling to find a way to easily define it. Some of the more universal symptoms – fatigue, post-exertional malaise, brain fog or memory loss – resemble what has previously been seen in other post-viral chronic illnesses, such as influenza or Epstein-Barr. But the sheer scale and heterogeneity of long Covid has made it challenging to study.

A number of different, compelling hypotheses have emerged

to try and describe the pathology underlying long Covid. Some have suggested the condition is caused by a persistent viral reservoir of SARS-CoV-2 viruses, the microbes that cause Covid, hiding out somewhere in the body. Others argue the acute illness triggers a chain reaction of immune dysregulation, which ultimately leads to persistent chronic symptoms. It's also been proposed that SARS-CoV-2 could alter one's gut microbiome in ways that cause broader systemic inflammation. And then there's all the ways Covid can impair normal functions of the brainstem and vagus nerve.

In October 2023 a study was published in the journal *Cell* that turned the world of long Covid research upside down. The study, led by a team from Perelman School of Medicine at the University of Pennsylvania, presented a kind of grand unifying hypothesis attempting to tie together all prior ideas surrounding long Covid.

The researchers first reviewed metabolite profiles from several previously published long Covid studies. A pattern quickly emerged. Those patients with long-term symptoms consistently showed lower levels of circulating serotonin. In fact, the pattern was so reliable the researchers could distinguish long Covid patients from fully recovered patients just by measuring plasma serotonin levels.

Subsequent animal tests revealed SARS-CoV-2 infections did indeed reduce circulating serotonin – so the researchers then wondered how this was happening. Because the vast majority of serotonin in our body is produced in the gastrointestinal tract, all attention turned to the gut.

Across a series of impressive animal and organoid experiments, the researchers discovered SARS-CoV-2 infections induced a kind of inflammatory response that disrupted the gut's ability to absorb the amino acid tryptophan. Without an effective source of tryptophan, the GI tract is unable to effectively produce serotonin, and this is potentially how serotonin depletion could be a defining trait of long Covid.

The final piece of the puzzle was understanding how this viral-induced serotonin depletion could lead to the neurocognitive problems commonly seen in long Covid. After all, serotonin produced in the gut does not cross the blood-brain barrier. While circulating serotonin levels looked to be directly depleted by viral inflammation, levels of serotonin in the brain remained unaffected.

Here the researchers turned to the vagus nerve – a crucial communication superhighway that travels from the brainstem to the gut. It was discovered serotonin depletion in the gut dampened vagal nerve signalling to the brain, specifically the hippocampus. When vagal nerve activity was reduced, hippocampal neuron activity declined, and this led to cognitive impairments such as memory problems.

The masterstroke in the new research was its exploration into how serotonin interacts with vagal nerve neurons. Elaborate animal and cell tests revealed serotonin signalling via 5HT3 receptors on vagal nerve neurons was responsible for this whole chain reaction. And, perhaps most significantly, when 5HT3 receptors on the vagal nerve were artificially stimulated with a drug, animals suffering long Covid-like impairments showed notable cognitive improvements.

What psychedelic compound is known to stimulate 5HT3 receptors? Ibogaine.

No magic bullet

Ash was fastidious about maintaining a treatment diary. 'Start low, go slow,' she'd say in reference to bringing any new compound into her larger regime. In the merry-go-round of self-experiments, she looked to a homegrown iboga tincture. Maybe, in low doses, it could help give her the motivation to exercise, she thought.

'Some people are extremely responsive to iboga,' Ash said. 'It has adverse effects when I take a drop of the homemade tincture without diluting it down.'

These minidoses of iboga did help Ash start exercising regularly again. They weren't traditional microdoses but something closer to a psychedelic dose. The psychological boost from this motivational bounce sent positive ripples throughout the rest of her life, but the iboga was no panacea. The brain fog and sense of disconnectedness was still devastating. Ash would regularly spend hours just staring into space.

Some time passed and a colleague offered Ash a magic mushroom extract. It was an unpleasant-tasting homemade concoction but being the psychonaut scientist she is, Ash gave it a shot. She took a minidose of the extract. The result was a deep healing sleep and Ash was excited. She took another small dose the very next day but it frustratingly did nothing.

She called her colleague with the news about the inconsistent results. 'No, no, no,' the colleague said. 'You can only take this once or twice a week.'

'So I eyeballed a dose twice a week,' Ash said. 'And the mental clarity started to come back. And they kind of feed off each other. So if you've got enough physical energy to do something, then you can exercise, you can go outside, you can have a shower, wash your clothes, you can hang them up. And you can do that best if you've had a deep night's sleep, which I hadn't had for a long time.'

The magic mushroom extract helped Ash regain a substantial volume of mental clarity and connectedness. Like any chronic disease, improvements were gradual with frequent relapses. Two steps forward, one step back, as they say.

'[But there] was still a feeling of derealisation, depersonalisation,' Ash said. 'Things aren't connected to each other, like being hungry and having food in the house are two completely separate issues. [And] it can take you an hour to connect.'

So Ash began experimenting with DMT. Just once or twice a month in a smokable blend commonly known as changa. Much like Ash's prior psychedelic experiments, the DMT was imbibed in low doses.

'I wasn't out to discover the meaning of life. I just wanted to get far enough in that I could access reality on a step-by-step basis. You can't afford to be that far gone if you're not processing things properly. If you can't remember how to put a meal together, the last thing you want to be is crawling around the floor.'

The psilocybiome

'The gut microbial system, at the interface between the individual and the environment, is important for healthy homeostasis,' explained John Kelly, a psychiatrist and neuroscientist from Trinity College Dublin. 'Gut microbes communicate with the brain via the gut–brain axis, including tryptophan–kynurenine, immune, hypothalamic pituitary adrenal axis, and vagus nerve pathways.'

Kelly has worked on several psychedelic therapy clinical trials, including last year's landmark phase 2 human study exploring the effect of single psilocybin doses on treatment-resistant depression. However, one of his personal areas of interest looks to bring together two different nascent scientific fields – psychedelic medicine and the microbiome.

In late 2022 Kelly and colleagues published a curious paper that proposed a novel hypothesis. If the trillions of microbes in our gut are being found to play a role in everything from psychiatric disorders to autoimmune disease, then what effect do these biofeedback signals have on the therapeutic outcomes of psychedelic medicine? Here Kelly and colleagues coined the term 'psilocybiome,' a combination of the words psilocybin and microbiome.

'The term "psilocybiome" is still somewhat conceptual at this point, but broadly refers to the interaction between psychedelics and the microbiota–gut–brain axis,' says Kelly. 'In other words, the psilocybiome comprises the reciprocal host–microbiota–psychedelic interactions and serves as an example of systems interconnectivity.'

Almost all questions surrounding the effect of psychedelics on gut–brain communications remain unanswered. In fact, both worlds of psychedelic and microbiome science are filled with countless unknowns right now. So much so that most reasonable scientists are incredibly cautious when talking about these things.

'It is established that serotonergic psychedelics have immuno-modulator and anti-inflammatory properties,' Kelly said. But the precise immune-signalling pathways and their relationship to the vagus nerve as related to potential therapeutic effects 'have yet to be fully understood'.

Kelly says the serotonin hypothesis for long Covid is a commendable piece of translational research. It points to fascinating directions for future research into the effects of serotonergic psychedelics on the gut.

The traditional psychoactive brew ayahuasca, which contains the powerful psychedelic DMT, for example, is well known for its unpleasant physical effects. Nausea and vomiting are consistently reported to precede the brew's hallucinogenic onset.

Scientists have suggested the purging associated with ayahuasca is the result of a flush of serotonin released by the gut. This serotonin release then stimulates the vagus nerve, causing the acute vomiting that accompanies the experience.

Attila Szabo, from the University of Oslo, is perhaps one of the world's top experts on the immune-modulating effects of psychedelics. He says the recently published serotonin hypothesis for long Covid is 'outstanding' and 'exciting', but he also warns against self-experimentation with psychedelics, even for post-acute sequelae of Covid-19, or PASC, another name for long Covid.

'I can see the beauty of a hypothesis on testing the potential therapeutic effects of, for example, microdosing classic psychedelics to tackle the pathophysiology of decreased systemic serotonin in PASC. However, we desperately need more human clinical studies, before any recommendations could be formulated here,' Szabo explained in an email to Salon.

According to Szabo, we have very little understanding of the systemic immune effects of psychedelics in humans. And in some cases psychedelics could hypothetically be harmful.

'This is something that everybody needs to understand: until the first reliable human clinical data is getting published, self-experimenting with psychedelic substances to treat inflammation, autoimmune disorders, symptoms of PASC, etc. is clearly not advised. Since classic psychedelics can also potentially cause system-level immunosuppression (both short and long-term, depending on drug use habits and other factors), avoiding them in the chronic phase of a disease seems to be wise,' Szabo stressed.

Kelly agrees, even indicating a need for further preclinical work to better home in on the complex interactions between psychedelics, the immune system and the psilocybiome. One compelling point Kelly makes is that psychedelic medicine generally focuses on the acute effects of a small handful of big doses. Other medicines, such as antidepressants, are taken daily for months, or even years, so their effects can be more consistent. Does one big dose of psilocybin lead to persistent benefits when thinking about conditions like long Covid, or will we need to explore longer-term low doses?

'It is important to consider the different durations of treatment between SSRIs or tryptophan supplementation and the serotonergic psychedelics,' Kelly explains. 'Clinical studies typically use between one and three doses of psychedelic in the context of therapy. As such, it will be interesting to figure out the extent to which one to three doses of psychedelic would be mediated by the pathways of the microbiota–gut–brain axis in post-viral conditions, such as long Covid and other immune-dysregulated conditions.'

A search for more scientific evidence

Saleena Subaiya is an assistant professor at Columbia University's Department of Psychiatry whose personal long Covid story started in early 2021. The condition forced them to quit their job as an emergency room physician in a busy New York hospital. Like many long Covid patients, they faced doctors who dismissed their condition as unrelated to Covid. They were put on numerous medical treatments, both traditional and non-traditional. Ultimately, Subaiya found their own way to live with long Covid. Diet and lifestyle changes were crucial, but to this day they still have good and bad days.

Over the last couple of years their interest in the potential for psychedelic medicines to treat long Covid led them to establish the first human clinical work directly investigating the relationship. Subaiya is currently running two human trials, one in the middle of recruitment and the other beginning in the new year. These initial trials are small pilot studies looking primarily at the effects of certain psychedelics on psychiatric and neurocognitive symptoms associated with long Covid.

Echoing Szabo and Kelly, Subaiya is well aware of how uncharted this territory is. Long Covid is a new illness, still being fully understood. Psychedelic medicine is a nascent field, with so much yet to be learned.

'You can understand how having a novel disease with a novel mechanism that we're still trying to figure out, as well as trying to use novel therapeutics, ones that have been illegal for many years and are still illegal, presents a challenge,' Subaiya explained in an interview with *Salon*. 'And so it's really, really important that we proceed very carefully. And we don't over-promise when the data is burgeoning.'

Subaiya says the recently proposed serotonin model for long Covid is exciting, and points to a potential way some psychedelics could help some patients. But from their experience, the heterogeneous nature of long Covid means this hypothesis is unlikely to encompass all presentations of the illness. They cited patients with long Covid–like symptoms from post-vaccine injury as a good example.

'So these patients present with a similar constellation of long Covid symptoms, but they don't actually have viral persistence,' they explained. 'And so this could be one theory that explains a portion of the patients that have chronic fatigue, brain fog, etc. But I'm not entirely sure that this presents the entire picture.'

One of Subaiya's biggest concerns for long Covid patients who may be experimenting with psychedelics is the possibility of their condition getting worse. Post-exertional malaise is a hallmark of long Covid for many patients. This is where cognitive, physical or emotional exertion causes a flare-up in disease symptoms.

'So we know that serotonergic hallucinogens can create a very, very intense and powerful psychological experience. When you're taxing the body, it can possibly even be detrimental to patients if, in fact, they're having a fearful experience. They're not adequately prepared. They don't have a provider to walk them through some of the bigger questions that can result after these experiences that can create ongoing stress. These are all really, really critical.'

'There is no quick fix'

'I can confidently say now I'm probably 90 per cent back,' Ash said. 'I can ride my motorcycle short distances. I'm not waking up with facial bruising in the morning. I can use scissors. I can memorise six digit strings of numbers. I still can't play guitar. I can't lift anything over my head. There's a bunch of stuff I can't do. But looking back on where I was in May 2020 it's a much more fabulous place.'

For Ash, psychedelics were crucial to her slow recovery, but they were also part of a larger therapeutic regime. A regime that included changes to diet, exercise, lifestyle. A regime that included countless experiments with other botanicals, herbs and vitamins.

Subaiya certainly believes in the potential for psychedelics to play a therapeutic role in the treatment of long Covid. However, they are still struck with so many questions that they urge people to not start self-experimenting.

What specific psychedelics could benefit long Covid patients? What particular populations of long Covid should be targeted? What kinds of doses help? What kinds of doses harm? These are all questions with no good answers right now.

'We're in a mental health crisis,' Subaiya said. 'We want a quick fix. And the answer is that there is none. We need a holistic model of care that takes into account the individual, their medical needs, their psychological needs, their unique cultural backgrounds, all of those things are so critical. Because of the hype around psychedelics, because there's no treatment for long Covid, we have a vulnerable population that is desperately seeking a treatment. And this concern that I have is that the data is just not there yet. And so, you know, it's really, really important to proceed cautiously.'

From Ash's perspective she is cautious to recommend anything specific to anyone. She recognises the privilege she had in terms of both having access to a community of scientists and plant experts, and the space to develop a personal treatment regime.

'I'm really lucky to be from that generation of psychonauts, who considers holistic medicine to be diet, exercise, lifestyle, self-reflection, documentation. Taking care with these things, you can't really separate anything out.'

And her advice to anyone with long Covid is to remove as many stresses from your life as you can, and be gentle on yourself.

'Pick one thing, start low, go slow, and monitor it. Whether it's diet, exercise, meditation, medication, or psychedelics.'

✱ *The consciousness question in the age of AI*, p. **76**
   *'Why would you find me attractive?': The body disorder that
      needs more attention*, p. **150**

CECILIA PAYNE-GAPOSCHKIN

Alicia Sometimes

*There is no joy more intense than that of coming upon a fact
that cannot be understood in terms of currently accepted ideas*

spent time measuring | | absorption lines | | in stellar spectra

     peered through a jeweller's loupe
     and poured through data etched
     into thousands of glass plates

calculated composition of stars | | mostly hydrogen then helium

Harvard College Observatory in the 1920s
(paid out of the director's equipment budget)

the dean of American astronomers wrote
to say her findings were *clearly impossible*
only to confirm a few years later it was fact

hydrogen, far more prevalent in the universe
than anyone had believed | | a million times more

radiant bearings in patchwork night
the lightest elements carrying weight

Cecilia became the first woman promoted || full professor
Harvard's Faculty of Arts and Sciences || department head

hydrogen || monarch of chemical
substances || atomic number one

this stairway of acuity
                making the bridge to new discoveries

                                          luminous

✱ *Poetic constellations – an exploded sonnet sequence*, p. **97**
   Origins – of the universe, of life, of species, of humanity, p. **104**

THE LAST KING ISLAND EMU DIED A STRANGER IN A FOREIGN LAND

Zoe Kean

It was 1805 and amongst the sweeping lawns of the Château de Malmaison in Paris a small, dark emu lounged in the sun. Black swans graced the lake and kangaroos dispersed as Empress Josephine, consort to Napoleon Bonaparte, would show guests her latest exotic treasures.

The emu was rare as it was the last of its kind – an endling.

This diminutive creature was a King Island emu, *Dromaius novaehollandiae minor*, a small, dark subspecies unique to the island it was named after.

'The mainland emu is around 30 to 45 kilos, and the King Island emu seems to have been closer to 20 kilos,' said David Hocking, curator of vertebrate zoology and palaeontology at the Tasmanian Museum and Art Gallery.

King Island was connected to the mainland and Tasmania during the last ice age and only became distinct when the earth warmed and sea levels rose around 14 000 years ago.

'That fairly short period of isolation was enough time for the King Island emu to dramatically change,' Mr Hocking said.

So how did this creature make it to an empress's garden? And what befell its brethren back home?

Voyages of discovery

The King Island emu was part of a remarkable scientific haul brought to France on board the vessels of the Baudin Expedition. This was a scientific and imperialist mission to the Great Southern Land led by Nicolas Baudin and commissioned by Bonaparte.

'They were trying to chart the missing part of Australia,' said Stephanie Parkyn, former ecologist and author of *Josephine's Garden*.

Their quest for knowledge went beyond charting and geography – at this time the natural sciences had prestige and France was keen to flex its scientific muscle.

'There was a rivalry with the English and nations were very proud of their scientific knowledge,' Dr Parkyn said.

The expedition left Le Havre, France, in 1800 with two ships, the *Géographe* and *Naturaliste*, and a whopping 22 scientists on board. This included young naturalist François Péron, who joined the expedition to escape a heartbreak, and who would go on to collect the King Island emu.

Discovery, disaster, death

When their ships reached King Island in December 1802, the French explorers found they were not the first sailors to drop anchor in its waters.

Ancient evidence shows Aboriginal use of King Island before colonisation, but it is believed the remote ecosystem was uninhabited by humans when non-Indigenous people first arrived. The first non-Indigenous person to clap eyes on King Island was a sealer, William Reed, who is believed to have sighted the island in 1799. The beaches were packed with lumbering, blubbery elephant seals, whose slaughter and processing for oil and skins became one of Tasmania's first boom-and-bust extractive industries, along with whaling.

By the time Peron arrived, the sealers had well and truly set up camp and the island was very different to today.

'There was a lot of native vegetation; it hadn't been cleared like it is now,' said Cathy Byrne, senior curator of zoology at the Tasmanian Museum and Art Gallery. 'There was a lot of very thick coastal heath ... and [emus] would push their way through the undergrowth.'

Peron's first glimpse of the emu was gory. He noted in his journal that they were 'hanging from a sort of butcher's hook' in a sealers cabin. The sealers hunted emus with dogs and welcomed the scientist.

In her Masters thesis on the emu, Stephanie Pfennigwerth highlights Peron's delight at being offered a soup made of a medley of wombat, emu and other mystery meats, which he described as a 'savory meal'. His notes about the emu show that their palatability was front of mind, writing: 'The flesh ... halfway (so to speak) between that of a turkey-cock and that of the young pig, is truly exquisite.'

One emu at least escaped his plate and was taken aboard the *Géographe*, destined to outlive its enthusiastically hunted kin that would disappear entirely by 1805.

The island's wombats, elephant seals and quolls were also eradicated.

Baudin's arc of horror

The emu arrived in Brittany, France, aboard the *Géographe* in March 1804, 15 months after it was loaded onto the ship. It had shared its hellish journey with a strange assortment of creatures, including the Kangaroo Island emu, kangaroos, black swans, frogs, tortoises and non-Australian species such as lions and mongooses.

So how did they survive the trip?

'Not all of them survived ... but Baudin insisted that his

lieutenants give up their cabins to the animals to try to keep them alive,' Dr Byrne said.

Baudin's journals note the struggle of keeping the creatures going, writing:

> Since the emus refused to eat, we fed them by force, opening their beaks and introducing pellets of rice mash into their stomachs. We gave them, and the sick kangaroo likewise, wine and sugar, and although I was very short of these things myself I shall be very happy to have gone without them for their sake if they can help in restoring them to health.

Baudin's own health was suffering by this point, as his observation continued to note: 'I had a worse bout of spotting blood than I had had before.'

The emus would survive the trip but Baudin succumbed to tuberculosis before arriving home. The ship was plagued by scurvy and dysentery, as well as tuberculosis, and many sailors and scientists died aboard. Only three of 22 scientists on the expedition made it back to France, with ten jumping ship early in Mauritius and the other nine falling to the hardship of travel.

Peron was the lone zoologist to return and was celebrated for collecting at least 100 000 preserved specimens, a vast wealth of flora and fauna prized in its day and still used by researchers today. The live animals were particularly prized by the Empress, who 'had a keen interest in the natural sciences', Dr Parkyn said. She said having Australian species in the garden would have been the 19th-century equivalent of having items brought back from Mars in your house.

'There's no way she could have travelled to see them herself,' Dr Parkyn said.

What remains?

Tasmania's relics of the King Island emu are limited to the contents of a few small boxes in research facilities. Dr Hocking cares for several of these specimens behind the scenes at the Tasmanian Museum and Arts Galleries collection.

'These are preserved fossilised bones that have come out of sand dunes,' he said.

The fossils are from animals that lived and died on the island and could range in age from just before the animal's extinction to thousands of years old.

The last King Island emu outlived the empress, dying in 1822. It underwent taxidermy and is held by the French National Museum of Natural History. One feather from the far-flung endling found its way home when it was gifted back to Tasmania by the French museum.

'You wonder if it knew what had happened to the rest of its kind,' Dr Hocking said, looking down at the delicate, double-stranded specimen.

It is cherished, and lies under a plastic sheet in its own beribboned box. The lonely feather of the last ever King Island emu is considered too precious for display.

✳ *Of moths and marsupials*, p. **58**
 The world's oldest story is flaking away.
  Can scientists protect it?, p. **111**

CALL OF THE LIAR

Carly Cassella

Down in a gully, beneath a canopy of tree ferns and towering mountain ash, I hear the echoing clacks and undulating whistles I've been searching for. It's taken a leech bite, a few thwacks to the face by branches, and a plunge into shin-deep mud on this crisp day in late May, but at last ecologist Alex Maisey has guided me to his home's greatest attraction: the superb lyrebird.

The creature is brown and pheasant-sized and nearly lost in a sea of green, save for that it sings and dances atop a scratched-up mound of dirt. I can tell by its lyre-shaped plume that it's a male. His shivering silver and brown tail feathers spread and fold over his face, almost like a human hand cups to share a secret. But this is no whisper. If a female were on the mound, he'd be screaming at her. As I watch through binoculars, Maisey translates the bird's best impressions: 'Grey butcherbird. Crimson rosella. Golden whistler. Magpie.' I recognise the roll of kookaburra laughter. My jaw drops when the bird mimics the whooshing sound of wings.

Superb lyrebirds, *Menura novaehollandiae*, are world famous for their near-perfect imitations. Human admirers come from far and wide to hear them here in Sherbrooke Forest, just outside of Melbourne, Australia, on the custodial lands of the Wurundjeri people. Each winter for more than 60 years, local enthusiasts like Maisey have hiked through the forest at dawn to keep tabs on the lyrebirds. One man from Melbourne was so obsessed with observing the species that he opted to live part-time in the forest within a massive hollowed-out log.

Of particular interest are male lyrebirds, who not only look showier but act that way, too. They have several different songs. One is a series of whistles, whoops, wails and warbles that, aside from a few bouts of embedded mimicry, is unique to the species. Other songs are used to court females and include up to 80 per cent mimicry. Lyrebirds can copy the sounds of more than 20 other species as well as the clack of beaks and the squeak of rubbing branches. They can even sing two overlapping bird songs at once.

Their outlandish performance is a classic example of an excessively complex trait or behaviour that was likely shaped through 'sexual selection'. Charles Darwin first introduced this theory in 1859 to explain why elaborate, seemingly non-adaptive characteristics, such as a peacock's tail, persist over generations. Virtually all birds make short calls to communicate. But according to Darwin, birdsong, which is more lyrical and complex, belongs almost exclusively to males, who use it to either attract mates or establish territory; the role of females is to silently judge whose melody is most appealing.

What started as a revolutionary idea in biology has, in the time since, closed many scientists' ears to competing narratives. Indeed, when biologist Anastasia Dalziell first started studying lyrebird songs in the late 2000s, she only focused on males as only male lyrebirds enjoyed a reputation for mimicry. Shortly after finishing her doctorate on lyrebird mimicry, however, the musician-turned-ornithologist realised her thesis needed revision. During a stint of fieldwork in the Blue Mountains, a forested landscape of peaks and valleys west of Sydney, Dalziell kept hearing female lyrebirds singing and mimicking.

People who live near lyrebirds had also heard females vocalise, especially when someone approached a nest. But no researcher had formally documented these instances in the scientific literature. Historically, scientists have dismissed female song as exceptional, inferior or functionless – an accident of sexual selection. In 2014, however, a global survey of songbirds revealed that lyrebirds are far from alone in their abilities: females sing in a whopping 71 per cent of species examined. Compared to other regions, females in Australia – where songbirds first evolved – are particularly vocal. And the lyrebird lineage is the oldest of them all, with fossils dating back more than 15 million years.

Eager to hear some of the most time-honoured female voices in the world, Dalziell, together with Justin Welbergen, set up a dedicated line of research as part of their program called 'Lyrebird Lab'. Ecologist Vicky Austin of Western Sydney University has led the ear-opening endeavour. Her findings suggest that female lyrebirds are just as skilled at singing as males, though they do so for different reasons. In the Blue Mountains, Austin has shown that females consistently whistle their own tunes, interlaced with imitations of other birds.

Her experiments in the wild are the first to suggest that lyrebirds use mimicry to protect the nest. For example, when a lyrebird mother feels threatened by a pied currawong, an egg eater with leering yellow eyes, she will often imitate the calls of larger predators, possibly to frighten or confuse, and draw the burglar away from the nest.

'We've only just unravelled the fact that female birds sing, but there's this whole other section now about mimicry that we're trying to get to the bottom of as well,' Austin says. 'If females aren't using mimicry to attract mates, then how does it evolve?'

For thousands of years, superb lyrebirds have drawn a human audience. Some Indigenous Australians know the bird as beleck-beleck or balangara and many describe it in their Dreamtime stories as the peaceful broker of the bush – the only animal able to communicate with all others. The first colonial settlers found the lyrebird's song disturbing and unnecessarily hunted males for their tail feathers, which fall out each year and can simply be picked off the forest floor. It took more than a century for the species' tune to become a cherished anthem for the new nation. In 1931, a male lyrebird from Sherbrooke was the first wild bird broadcast on Australian radio.

Even before recording sound was possible, tales of the lyrebird's song drew the attention of John Gould, the so-called father of Australian ornithology, as well as the first scientists to formally describe evolution, Alfred Russell Wallace and, of course, Darwin. None of these English naturalists could ultimately agree on why the lyrebird sang or mimicked to such excess. Gould did not accept the theory of natural selection, and Wallace thought a female must choose a trait or behaviour because it offered some concrete survival advantage to her offspring, not because she found it aesthetically appealing.

But where Wallace found the male lyrebird 'most conspicuous' and 'wonderful,' he described the female merely as 'unadorned'. Similar perspectives riddle scientific literature. In 1900, Australian-born naturalist Archibald James Campbell wrote, for example, 'that only the cock bird whistles and mocks in this magnificent style; the hen makes but feeble attempts'. Darwin, for his part, also hewed to the idea of male superiority. He used lyrebirds to illustrate the central idea of his seminal book, *The Descent of Man, and Selection in Relation to Sex*, which explores the struggle between males of various species 'for possession of the females'.

Pretty much ever since, feminist evolutionists, sometimes called Darwinian feminists, have been working on a more inclusive theory that doesn't assume males are naturally 'superior'. 'Social selection theory' is one way to level the playing field. Popularised by the biologist Mary Jane West-Eberhard in 1978, the theory has come to frame all animal interactions as forces that can shape species, not just those among potential mates or between males. In both sexes, the theory goes, competition and cooperation for food, territory, nesting sites and other essential resources can contribute to the development of elaborate traits.

It was a duet that initially inspired ornithologists to take another look at the social context of female song. When barred owls

(*Strix varia*) court, the male and female trade a series of alternating gurgles, hoots, caws and cackles that echo through the treetops like a troop of noisy monkeys. A behavioural ecologist named Karan Odom studied the duet in the late 2000s and began to wonder: if each sex has their own distinct part, could females of other species be singing too?

For more than a century, Western science had defined birdsong as the longer, lyrical and more elaborate vocalisations that male birds make during the breeding season. But what if that definition was too narrow? Odom joined with other like-minded scientists to look back in history and see if females were ever heard vocalising in complex ways, even if, at the time, that wasn't technically considered singing. She spent months poring through 16 large tomes detailing every described bird species.

'I was really surprised how much female song was documented, how much it seems to crop up, again and again,' Odom, an author of the 2014 survey, tells me. Across the songbird tree, female voices were clamouring to be heard. Odom and others went on to create the Female Bird Song Project, an online database that aims to systematically catalogue female birdsong globally.

Down Under, researchers at the Lyrebird Lab are doing their part. Austin, Dalziell and Welbergen are the first to place cameras and microphones in and around lyrebird nests. 'It's not all about the males. Rights for females!' Austin half-jokes, pumping an imaginary protest sign as we scramble through wet forest, across streams, over boulders and under fallen trees on a rainy day in September. We are hunting for female lyrebirds in the Blue Mountains, on the traditional lands of the Dharug and Gundungurra peoples. Suddenly, I hear a high-pitched whip-crack echo through the blue gums, closely followed by a double note: *Cooooowheep! Pew! Pew!* Austin identifies the sounds as a male and a female whipbird (*Psophodes olivaceus*). The coordination of their duet was so finely tuned that if I assumed males were the only singers, I'd have never heard the female's voice. Austin keeps her ears peeled but she guides us mainly with her eyes, surveying the forest floor for signs that a lyrebird has been scratching. With young to protect, females rely on silence and subterfuge while they forage for insects or build their nests. Finding them requires dedication and, like today, many unsuccessful outings. It took Austin months to be able to tell the difference between a lyrebird mimicking a goshawk and an actual goshawk – and that distinction is based more on location and behaviour than on any noticeable difference between the calls themselves.

'Look,' Austin tells me matter-of-factly when I ask why female song has been disregarded for so long, 'a big part of it is that males capture our attention. Males are pretty; they're loud; they're obnoxious; they're just in your face. They're beautiful, and I get it, but it's the subtlety of females that interests me. You have to be patient.'

Austin's years of persistence have paid off. Dressed in camouflage, she's spent hours crouched outside lyrebird nests, waiting to hear their morning whistles. Sometimes, she's noticed, they also sing while foraging. Similar to male lyrebirds, the female's whistle song is embedded with bouts of mimicry, although their imitations are not as repetitive and they tend to copy more predators. The reason for the whistling is unclear, but Austin suspects it involves competition for space and resources – and the ongoing need to signal one's presence to defend those assets. Female lyrebirds are aggressive with one another, destroying each others' nests and attacking each others' young. Mothers often build multiple nests throughout their territory, possibly as back-ups, decoys or territorial markers. Whistling a song every morning could be a safe and easy way to signal: no trespassing.

In one set of experiments, Austin challenged wild female lyrebirds as they left their nests, pulling back a sheet to reveal a stuffed fox, currawong or rosella. Each of these taxidermied intruders triggers a different response from lyrebirds. Females mostly ignore the innocuous rosella, a colourful parrot that poses no threat to lyrebirds or their offspring. But when a fox is present, females scream in a way that implies multiple voices, like an army of birds facing a threat. Lyrebirds might produce this disorienting alarm to fool a predator into thinking it's outnumbered or to recruit other animals to drive the threat away.

Once, Austin observed a mother respond to a currawong at her nest by calling in an ear-splitting mob of bell miners, a native species of honeyeater that swarms and chases native birds from food sources. Another of Austin's recordings shows a female lyrebird flapping and singing around a goshawk perched at her nest entrance. Three-quarters of her cries involve mimicry, mostly of smaller birds such as scrubwrens, king parrots and whipbirds. More detailed research needs to be done, but Austin is convinced that mimicry 'isn't just a pretty thing. The female is actually using it functionally.'

And why not? The stakes are high and lyrebird defences are limited. Each year the female lays just one egg. This chick takes six weeks to hatch and another month to leave the nest, which is unusually slow for a bird. Meanwhile, males don't care for offspring at all, leaving females responsible for the species' survival.

'We've now got to consider that, in fact, mimicry was all about anti-predator function or about nests,' Dalziell explains. 'And then it got co-opted into a sexual signal.' Most research on vocal mimicry to date has focused only on males, but the few studies that involve both sexes suggest mimicry is used for more than just courtship. In 2017, researchers in the United Kingdom recorded common cuckoos (*Cuculus canorus*) imitating predatory sparrowhawks to scare other birds off of their nests. The cuckoos then snuck their own eggs into the mix and escaped parenting duties. In 2021, another team in China caught female great tits (*Parus major*) hissing like snakes when squirrels threatened their nests. In South Africa, scientists recently documented fork-tailed drongos

(*Dicrurus adsimilis*) mimicking meerkat calls that communicate a nearby threat. After the meerkats fled, the drongos stole their food.

Other times, it's not clear why birds imitate the sounds of other species. Scientists in North America recently discovered that female northern mockingbirds (*Mimus polyglottos*) also mimic like their famous male counterparts. Ornithologist Christine Stracey was placing cameras in and around mockingbird nests to study their reproductive behaviour when she noticed an audio feature on the device. She flipped the switch and opened her ears to a whole new world of bird sounds. Stracey shared her recordings with Dave Gammon, a biologist who has studied the species for more than a decade. Sure enough, he confirmed the females were imitating other species.

Male mockingbirds mimic to court and define territory. But researchers have yet to discover why females are mimicking. Initial findings suggest that female mockingbirds might even expose their nest to predators when they vocalise. So why do it? Gammon thinks it could be a byproduct of the species' innate song-learning abilities. Stealing other songs could be an easy way to expand a repertoire, or it could even be a form of play. 'I think of [mimicry] as part of the package that got dragged along with [song],' Gammon tells me.

Even without a clear purpose, it's possible that an extreme trait like mimicry can persist through generations, via genes or the way a brain is wired – as long as it's not significantly detrimental. In other cases it really might be about beauty. As evolutionary biologist Richard Prum has pointed out, sexual selection isn't just about male prowess. It also posits that females powerfully sculpt the evolution of their species simply by choosing mates – what Prum calls 'Darwin's really dangerous idea'. Elaborate behaviours like mimicry don't have to be useful to exist, Prum says; they could be 'merely attractive'.

The idea that mimicry can have multiple origins is supported by hard data. Unlike birdsong, which appears to have arisen in a common ancestor and spread widely through the songbird tree, genetic studies suggest vocal mimicry evolved independently on

237 separate occasions and then disappeared in at least 52 of those cases.

Even within the same species, the use of mimicry is highly variable. For male lyrebirds, learning to vocalise seems to be open-ended, meaning that they continue to accrue and improve their repertoire throughout life. That's probably what allows them to mimic fire alarms and crying babies when housed at a zoo. But among females, initial studies suggest the best mimics are not necessarily the oldest. Alex Maisey tells me that one young lyrebird in Sherbrooke was so vocal, she was known to locals as the 'singing female'.

Compared to those in the Blue Mountains, most female lyrebirds in Sherbrooke do not sing consistently. Maisey has spent many a morning huddling near nests in the predawn to confirm what Austin was hearing a state away – but he was met with silence. Perhaps in a smaller forest, where lyrebirds and their predators are more concentrated, singing near the nest is simply too dangerous. When a chick is around, Austin says even females in the Blue Mountains are quieter than normal, most likely for fear of giving away their presence to predators.

In such moments of silence, it's worth remembering: if a female bird isn't singing, it doesn't mean she can't.

The easiest female lyrebirds to find in the Blue Mountains nest near the busy boardwalks of a place called Scenic World, a former coal mine at the base of a sheer sandstone cliff that's now a tourist attraction. To reach the forested valley, Austin and I ride the world's steepest passenger railway. We have just stepped off an established path when she holds up a finger. 'Do you hear that?' It's a note of goshawk. I look down to see a brown bird with a small, sleek head, reddish wings and a bent tail, its spindly leg frozen midway through pulling up a root. 'They look like dinosaurs,' laughs Austin, as the bird dashes behind a tree. We sit and watch the female forage, the quiet broken only by the occasional trail runner shouting a cheery g'day, oblivious to the bird in the bush.

Austin calls this lyrebird Diana. Her ears are now so tuned to females, she can identify individuals by voice alone and can discriminate between regional 'accents'. 'This is how we discover things,' Austin says. 'It's just humans being interested in the world around them.'

By the look of Diana's tail, she's been cooped up in her nest for weeks. In Austin's time knowing her, she has raised two chicks to maturity, which apparently is 'pretty good'. One year, though, a tabby cat – likely from a neighbourhood up on the cliffs – stole Diana's chick a day before it was due to fledge. But lyrebird mothers, perhaps, don't take losses like that lying down. 'There are a couple of female lyrebirds, one here and one in the gully just over, that mimic cats,' Austin says.

She's never heard a male meow, but she's never searched for it, either.

SCIENCE IN THE BALANCE

Jacinta Bowler

University, for many, is the first real taste of freedom. Behind the O-Week parties, the huge lecture halls and soon-to-be-friendships is an almost unbridled hopefulness for the years to come. This was the case for Suzanna Reece.*

When she first arrived at uni, she was excited, stressed and a bit overwhelmed. 'I really liked science, and I really liked animals,' she says. 'I thought I was going to become a veterinarian.'

That changed after starting a biology degree. Like many students studying STEM, it dawned on her that research might be her calling: 'I had a really amazing group of lecturers and tutors and academics,' Suzanna says.

After just one year at university she realised that research 'is what I wanted to do with my life'. She was understandably optimistic about what the future held. 'Stupidly so,' she tells me. 'It just looked so great on paper.'

There's a well-trodden path into research. It starts with a year-long honours extension of an undergraduate degree, followed by a doctorate (also known as a PhD): three or so years of intense but rewarding work, producing 'significant and original' research. If honours is dipping a toe into the sea of research, a PhD is jumping in fully clothed.

Around 10 000 people graduate with a PhD in Australia every year; cumulatively, 185 000 people had a PhD in Australia as of

2021. Most graduates want to be a researcher – a doctorate is the first rung on a career ladder to become a lecturer or professor. But in universities, these jobs are slim.

In 2021, there were 47 000 full-time and fractional academics across the country, just 15 000 of whom were professors or associate professors. And it's not just that career progression is difficult. Academics on all levels are struggling with funding constraints, difficult and sometimes toxic work environments, a lack of job security and more, often with little support from their institutions.

Universities, individual scientists and research are all suffering as a result. So what are we losing as a nation when scientists leave academia? Has it always been this way? And what are the solutions?

Not many people move further than Cedric Courtois to start a new life, even in the world of academia. He shifted from Belgium to Brisbane – almost halfway around the world.

'He arrived in Australia in March of 2020,' says Zoe MacLean, Cedric's friend and colleague. 'Seven days later, we're in lockdown.'

The University of Queensland (UQ) wanted Cedric for a reason. He was a rising star in his field of media research and algorithms. His work used data science to ask important questions about how we use technology and how it uses us.

He'd graduated with his PhD in 2012, and only seven years later was an associate professor at KU Leuven in Brussels. At UQ, he became a senior lecturer in the School of Communication and Arts.

'He was a really interesting humanities researcher because he had come up through the social sciences,' MacLean tells me. 'He had a very analytical approach to doing communications research. He was very quantitative, and he loved mixing methods across disciplines.'

Cedric also focused on the online: social media as communities, online dating, googled health queries. And when he wasn't working, he was cycling, 'He had at least one bike chained up at the university and then a couple more at home,' says MacLean. She laughs as she remembers the time he told her he needed a bigger apartment because one room was 'entirely filled with bicycles'.

Locked down, far from family and scrambling to set up a new course at a new university in a new country – those first few months in Brisbane couldn't have been easy. It wasn't going to get better.

Once students make it through the cold shock of a PhD, their next rung on the ladder is a postdoctoral researcher ('postdoc').

This is a position under a supervisor, usually a laboratory head, where the postdoc conducts research for the lab. PhD positions are relatively plentiful. The government provides funds to pay PhD students about $30 000 a year under a system called the research training program (RTP), so they're cheap for universities and labs. This is big business – overall, in 2024, Australian universities will receive almost $1.2 billion for the RTP.

Postdoc positions, on the other hand, are funded by the lab heads directly, usually as part of a larger grant to the whole lab. Postdoc positions are therefore scarce and fought for, despite usually being short-term or semester-to-semester contracts. Postdocs supplement these precarious, underpaid positions by university tutoring or lecturing, or by winning small grants and awards.

Moving further up the academic career ladder provides more responsibility, but not a lot more stability. The idea of tenure – a permanent position within a university that you can't be booted out of – died in the '90s.

'While tenure is a lovely idea, no one actually has that sort of job security any longer,' says Jess Harris, an associate professor at the University of Newcastle, who researches university education. 'No one's safe in this scenario ... Even in some of the big [Group of Eight] universities, entire departments are being dismantled.'

Harris interviews academics for her research and has come to understand the breadth of the problems they're facing.

'Not being able to get a mortgage has been a common theme, particularly for sessional staff. They're employed from semester to semester, which means that they're not seen as having full-time work. International researchers [can be] unable to get permanent residency in Australia, because they're not employed full-time.'

It wasn't always quite this dire, but historically, funding in academia wasn't ever easy or perfect.

Early in the 20th century, universities were funded by state governments, student fees and endowments; research was treated as a hobby for academics to undertake in their spare time. Funding for equipment was regularly provided by an academic's richer friends. The main job was to train students.

More funding for research came during wartime, with scientists like Ruby Payne-Scott and Joseph Pawsey undertaking radio astronomy research in the 1940s only because it was also helpful for the development of better radar. CSIRO's precursor – the Advisory Council of Science and Industry – was established during World War I and regularly undertook classified defence research until it became the CSIRO in 1949.

In a 1971 paper in the journal *Minerva*, political scientist Sol Encel wrote that the academic system evolved out of a combination of the federal government wanting more involvement in the state-based university system, the Cold War and the 'pursuit of rapid economic growth'. He notes that in the mid-1960s, the Australian Research Grants Committee – the precursor to the Australian Research Council (ARC) – was launched. Funds formerly given to universities to be divvied up among researchers would instead by distributed by application to the committee.

According to Encel, between 1966 and 1969, 58 per cent of applications submitted were granted. But by 1983, this had dropped to around a quarter of proposed projects being funded: 514 out of

2000. This just never got better. In 2005, one third of submissions to the ARC discovery program succeeded. In 2023, only one in five submissions were funded (16.25 per cent). The majority of grant applications aren't of poor quality – instead, the pool of money is getting smaller, while the expense and number of applications keep increasing. In 2023, for instance, the ARC received 4100 grant applications.

'It certainly has gotten worse over time,' says Harris. '[Older] colleagues that I've worked with over the last 20 years ... have said that they wouldn't want to start in academia in this kind of context – in this current climate.'

The grant system makes or breaks academic careers. Whole labs of people can become unemployed overnight if a grant application is rejected; projects that have had millions sunk into them can be canned halfway.

Applying for grants is time-consuming, and takes effort away from research. In 2016, the results of a survey of over 2000 academics was published in the *Australian Universities' Review*.

The survey looked at research workload allocation, and found that the average time it took to prepare just one 'nationally competitive research grant' was 120 hours. That doesn't mean that such applications are approved – just that it takes that much time to produce a competitive submission to a grant body like the ARC.

In 2013, the year British physicist Peter Higgs received the Nobel Prize, he told the *Guardian* that he doubted he would have been able to come up with the Higgs mechanism in today's academic institutions. 'It's difficult to imagine how I would ever have enough peace and quiet in the present sort of climate to do what I did in 1964,' he said at the time.

Suzanna had not just found her calling in biology, she excelled in it. Buoyed by those around her, she got high grades and sailed through honours, graduating with a first-class degree. But she'd begun to notice that her supervisor was pretty 'hands off', and it was instead the postdocs who would work with the honours and PhD students. With more and more pressure on senior academics – the grant writing, paper publishing, undergraduate teaching and more – this wasn't particularly unusual.

'I was largely left to my own devices,' she tells me.

When I ask when it all started to go wrong, she pauses for much too long: 'First year, PhD.'

Her PhD was an extension of her honours project, developing new drugs for use in animals.

'These are people's pets who are dying,' Suzanna says. 'I could see very clear outcomes.'

But it didn't go as she planned.

'My supervisor doesn't have a malicious bone in his body,' she begins. '[But] he was getting very distracted with grant writing. He was always trying to do [more] publications. He was always travelling to conferences ... He seemed sometimes more interested [in] and distracted by the side parts than what was actually going on [in his lab].'

According to Suzanna, what was happening in his lab was questionable science. Her team was testing a drug as it was moved from laboratory testing to animal testing. Normally a drug would go through rigorous safety testing in a lab – like toxicity testing on cells – before being tested on a live animal like a mouse. This is to ensure that live animals aren't unnecessarily subjected to pain. According to Suzanna, the lab skipped this step due to time constraints, instead relying on historical data from testing in another species. At this point, her speech slows. Each word is deliberate; it's clearly a hard tale to tell.

'I had some concerns about safety,' she says. 'I felt we were jumping the gun, and I was reassured by everyone around me that it was safe at the concentration we were using.'

When I ask her if it was safe, her voice cracks.

'None of the people who told me it was fine were the ones who had to see those mice suffer every day,' she says.

The mice ended up with unexpected complications and some were euthanised before the end of the experiment. Scientifically, 'there was nothing we could gain from it.'

When professors and lab heads are trying to keep their careers afloat, PhDs and early career researchers are left adrift.

In 2019 and 2022, a team led by the Queensland University of Technology's Kate Christian surveyed 500 early career researchers to analyse how the sector was changing. Published on pre-print server bioRxiv, their paper found that during the survey period, job satisfaction declined to 57 per cent (from 62 per cent), workload concerns increased to 60 per cent (from 48 per cent) and around half those questioned experienced bullying. Over three quarters think that 'now is a poor time to commence a research career'; this increased from 65 per cent in 2019.

When you're at the bottom of the academic pecking order, there's limited power to speak out if something goes wrong. If you tell someone about dodgy research, would your supervisor still let you finish your PhD? If you speak out about workplace bullying, will you be labelled a troublemaker? If you're only being paid $30 000 a year, can you really afford not to take whatever work comes up, even if it's unpaid for most of the year?

By 2022, Queensland had no more lockdowns, students were returning to campus and Cedric Courtois had begun to settle into life at UQ. That March, he hosted a webinar for the Language Technology and Data Analysis Laboratory about digital archives. From the video it's clear he's a great public speaker. His Belgian accent is slight; his hands are constantly moving in and out of frame.

'The impression that I got from him about academia was that he loved it,' says his colleague MacLean. 'He really enjoyed doing good quality research work. It was the red tape, the admin side of things, that frustrated him. He just wanted to research and teach and there was all of this extra stuff that he had to deal with.'

Cedric's teaching work was impressive. He developed a first-year course called Data and Society that looked at everything from censorship to online abuse and algorithms.

'He was the architect – the heart and soul of that course; it was incredibly insightful,' says Izrin Ariff, one of the tutors.

There's a line in Kate Christian's 2022 research paper that I keep coming back to: 'Perhaps conditions could be tolerable if the ecosystem were yielding well-trained scientists and high-quality science. Unfortunately, there are signs of poor supervision and high rates of questionable research practices.'

The paper found that more than 47 per cent of the early career researchers surveyed had been impacted by questionable research practices – things like excluding data points from papers, publishing negative results or carelessly assessing a colleague's work.

A Dutch study published in 2022 in PLOS One found that more than half of the 6800 researchers surveyed undertook questionable research practices. Worse, 4.2 per cent of participants admitted to falsification – that is, manipulating or omitting data – while 4.3 per cent admitted to fabrication.

In 2023, online news site Cybernews reported that a number of peer-reviewed papers had been written using generative language models like ChatGPT, but had been caught out because they forgot to remove the precursor words 'as an AI language model' before publishing.

An analysis published in *Nature* late last year found that, in 2023, the number of scientific papers retracted exceeded 10 000 for the first time. In 2013 there were just over 1000 retractions, but by

2022, the figure topped 4000 then jumped to more than 10 000 last year.

'The level of publishing of fraudulent papers is creating serious problems for science,' Oxford University psychology professor Dorothy Bishop told journalist Robin McKie in the *Guardian*. 'In many fields it is becoming difficult to build up a cumulative approach to a subject, because we lack a solid foundation of trustworthy findings.'

Retraction Watch – a database and blog highlighting retracted academic papers – has hundreds of plagiarism allegations. Just one example is from August 2022, when Nicki Tiffin, a researcher in bioinformatics, discovered she'd published a new study in the *European Journal of Biomedical Informatics* – except she'd never submitted to that journal. Over a year later, it still hadn't been taken down by the journal.

Staying in academia is a struggle, leading some people to undertake questionable research practices – and others to just leave. Some head overseas. Some move into industry. Others quit and end up working in completely new fields, taking all that knowledge with them.

Some, we just lose.

In September 2022, both MacLean and Ariff noticed something was off with Cedric.

'He told me that he had been suffering from depression,' says MacLean.

Cedric added that he didn't like the drugs his doctor prescribed, because they were messing with his sleep.

'He had replied to one of my emails at three or four in the morning … and I remember thinking to myself, "Mate, burning the midnight oil there, geez,"' says Ariff – but it was marking season, a stressful time. Probably nothing.

Three days after the late-night email, Cedric died by suicide.

As soon as the university knew, tutors were instructed not to tell students in Cedric's Data and Society course, so as not to upset them so close to the exams.

'We told students that there had been a change – this is your new course coordinator, here's your new tutors' details, but we didn't tell them why,' says MacLean. 'We didn't tell them that he died at all.'

Both MacLean and Ariff believe that the students still haven't been told what really happened.

Some of Cedric's teaching materials are still being used in tutorials and lectures, but he's absent from the UQ website. Sometime after his death, his academic profile was quietly removed.

We don't know the factors leading up to his death; more than a year later, there have been no real answers.

I interviewed a dozen academics in the course of reporting this story, and ended each interview by asking what they would do if they were vice chancellor. Most asked for simple changes – more secure work, better training for supervisors and more support from the university if something goes wrong.

'I'd implement a practice of supervisors actually supervising ... It's currently not built into their workloads,' says Christian. 'I would also be encouraging more professional development and more career advice for the early career researchers who are not going to be able to stay in academia but who ought to be prepared for life elsewhere.'

'I just wish that there was a panacea – that we had an answer for all of it,' laughs Jess Harris.

Most importantly, though, interviewees suggest we need to better support academics to do the research they want to do.

'People come out of their PhDs really highly trained to be very good at this specific little [topic] – and then we waste them,' says Christian. 'If we want ourselves to be taken seriously as a clever country, then it's time that politicians [quit] saying, "Oh, we need more people in science." We've got plenty of people in science, we just don't know how to keep them.'

Optimistic students still surge into universities around the country every year, and many end up washed up in the system that is supposed to support them to uncover the next Higgs boson, or design the next life-giving cancer drug. When the system leads them to step out of academia, many young researchers mourn the loss of their dream.

Suzanna had a run of bad health and bad luck that led her to put her PhD on hold. She's since worked in laboratory technician jobs, roles which don't require a PhD qualification. She says that if she knew back then all she knows now, 'I probably wouldn't have bothered with a PhD. I would have just gone into another generic science job, like the ones I work now. You're working the same job you could have started five years earlier, and you'd be five years further into your career.'

But she says she still loves her research, and still wants to finish her PhD so she can get back into academia.

'Sunk-cost fallacy,' she jokes. 'Or I'm a glutton for punishment.'

She's still trying to finalise her last chapters to resubmit her PhD. She's aiming to become a research assistant, or get a postdoc, but she is well aware that the going will be tough. Despite everything she's been through, she's still hopeful she can contribute to the larger science cause and make a difference to the academic system.

'I want to go back because I realised you have to be on the boat to right the ship,' she says.

'If every single person who actually cares leaves, what's left?'

* Names have been changed.

✱ *Doing drugs differently: For public health, not profit*, p. **162**
Indigenous science must be a standalone national science priority, p. **175**

EVERYTHING STARTS WITH THE SEED

Natalie Parletta

When Dr Sognigbé N'Danikou was a small boy, his grandmother cooked meals with yantoto, a wild green that grew on his mother's farm in Ouesse, a small village in Benin, West Africa, about 250 kilometres north of the nation's capital, Porto-Novo.

'At that time, I fell in love with this indigenous vegetable,' he tells me, a warm smile transforming his serious demeanor.

Years later, N'Danikou attended university to study agriculture in Benin's seaside town, Cotonou. But he couldn't find his favourite green, which he likens to a type of lettuce but more nutritious. 'I couldn't enjoy them anymore, and that is an indication of how traditional knowledge can be lost; this connection with my mother and the stories she told about these crops.'

Today, N'Danikou is a scientist focused on traditional vegetables conservation and utilisation as the African genebank manager for the World Vegetable Centre – or WorldVeg, as it's affectionately known – based in Shanhua, Taiwan. Since 1973, this non-profit organisation has amassed the world's largest collection of vegetable germplasm – live seeds or other viable plant parts – and their genetic profiles, which are stored in Taiwan and Tanzania.

The centre collaborates with public and private research institutes to collect, conserve and identify important traits of vegetable germplasm to help increase farmers' productivity and livelihoods, alleviate poverty and malnutrition in developing countries, and mitigate the environmental and health impacts of pesticides and synthetic fertilisers.

'We say everything starts with the seed,' says Dr Gabriel Rugalema, WorldVeg's regional director for operations in eastern and southern Africa (who introduced himself to me with his perpetual cheeky grin as 'Gabriel the archangel').

Diverse and inclusive

I visited WorldVeg in November 2023 for their 50th anniversary celebrations. As our airport bus pulls into the hub of its sprawling 216 hectares, an oasis formerly covered by sugar cane, a wiry man in a business suit is pedalling furiously across the campus on a pushbike.

'There's Marco!' says my companion, Cathy Reade from Australia's Crawford Fund – a not-for-profit focused on food security. I learnt that Dr Marco Wopereis, an agronomist, is the centre's Director General. A couple of days later I see him boogying down in jeans and a WorldVeg t-shirt in an African dancing demonstration during a public open day; in the days before he hosts scientific talks and government officials for the official celebrations on stages elaborately decorated with vegetables, kicked off by guttural, tribal drumming that feels like it erupts from the Earth.

In my room is an invitation to join board members and scientists for dinner, hosted by Wopereis and his energetic wife, Myra. Their house is a short walk from our accommodation, in a mini village of red-roofed houses set among stoic trees with large branches that form a welcoming green canopy. Wopereis later tells me they'd planted another 9000 trees on the campus a couple of years ago. Walking around, I notice that the landscaping is like a botanical garden with multitudes of familiar and exotic edible plants growing in between the lawns, trees and buildings, all tagged with their common and scientific names.

At the dinner, tables decorate the Wopereis's lawn with a banquet created from an array of – you guessed it – vegetables. Myra enthusiastically introduces the food, highlighting an exotic dish made from a white edible Asian flower, *Sesbania grandiflora*. Other delicacies include grilled eggplant, grilled squash and red pesto with pasta. After the feast, Wopereis gives a carefully prepared speech in which he mentions everyone there, from dignitaries to people who helped with preparations, and even me. I came to realise that this inclusivity is a central thread in the tapestry of the organisation's extensive networks.

Scientists travelled from regions all over the world to attend the celebrations, including Africa, India, Thailand, Germany, Japan, the Philippines, Korea, America, Britain and Australia. The accents wafting around, which also include French, Austrian and Dutch, reflect the diversity that is at the core of their united vision.

'If you lack diversity, you lack options,' says Dr Roland Schafleitner, who leads the centre's biotechnology/molecular breeding group and the Vegetable Diversity and Improvement Flagship. 'And when something happens, you lose even more options until you have nothing left. So, vegetable diversity is crucial ... [But] we risk losing diversity on all levels, and that's also with vegetables.'

Spreading the risk

One of the many benefits of diversification is that if a crop fails, you're not putting all your tomatoes in one basket, as it were; it can be supplemented by high-value vegetable crops such as mungbeans (*Vigna radiata*) or amaranth (*Amaranthus spp.*) to help spread the risk. Yet, as food systems become more centralised and people's tastes become more uniform, we are at risk of losing this diversity.

'We eat tomato, cucumber, some beans and carrots, maybe eggplant and broccoli,' says Delphine Larrousse, regional manager for East and South-east Asia, in her animated French accent. 'But about 1100 vegetable species are recognised worldwide – there is a massive disconnect.'

Importantly, diversity doesn't just apply to different types of vegetables – it's also a critical feature within species. With fewer genes in the pool, hybrid vigour is diminished, leaving plants weak and vulnerable. Diversity helps breeders develop improved varieties of vegetable species that are resistant to stresses such as pests, diseases, drought, floods and heat, and that can produce more nutritious crops with higher yields, longer shelf life, shorter harvesting times and desirable traits such as taste, size and colour.

Bitter gourd (*Momordica charantia*), for instance, is a highly nutritious type of squash that feeds at least a billion people in Asia, and the annual seed market is valued at $26.4 million. However, hybrid crossing of elite lines to fill niche markets has diluted its genetic diversity, as confirmed via molecular analysis. That has made it vulnerable to biotic (living) and abiotic (non-living) stresses and impacted long-term yield. WorldVeg scientists are now actively introducing diversity to breed new, improved lines and rescue the market.

Diversity also safeguards against future challenges. 'The issue is that you don't know yet what kind of traits you need in five years from now,' says genebank manager Dr Maarten van Zonneveld in his softly spoken Dutch accent, 'and the other issue is that you don't know which variety has the traits you want.'

In recognition of this, the genebank's collection includes about 8000 different accessions of mungbeans alone – around 10 000 including its wild relatives. (The UN's Food and Agriculture Organisation defines an accession as 'A distinct, uniquely identifiable sample of seeds representing a cultivar, breeding line or a population, which is maintained in storage for conservation and use.') There are more than 12 000 soybean (*Glycine max*) accessions, 8000 tomatoes (*Solanum lycopersicum*) and 6000 chilli peppers (*Capsicum annuum*). Ethiopian eggplant (*Solanum aethiopicum*) seeds have topped 4000, 'but there are so many more varieties out there,' says van Zonneveld.

It's hard for most of us to imagine thousands of different types of chillis, eggplant or tomato. But vegetable species consist of a wide range of varieties with 'an astonishing diversity of forms, tastes and colours, adapted to myriad environments and pests and diseases,' according to van Zonneveld. 'This is what we conserve in our genebank.'

Other lines include amaranth, okra (*Abelmoschus esculentus*), jute mallow (*Corchorus olitorius*), cowpea (*Vigna unguiculata*), Ethiopian mustard (*Brassica carinata*) and bitter gourd. Van Zonneveld worries that they only have about 4400 accessions of cucurbits (pumpkins, squash and gourds). 'The luffa [*Luffa spp.*], all these cucurbits, let's collect them before they get lost.' His personal dream is for the vegetable collection across global genebanks to grow from around 200 000 to 300 000 accessions by 2030.

'It's just a number, no, but something to aim for.'

Van Zonneveld's favourite seed rescue was a student's redis-covery of an endangered cowpea and mungbean relative named *Vigna keraudrenii*, endemic to Madagascar. It was initially found in 11 spots among the island's rich biodiversity in the 1990s and described by botanists David du Puy and Jean-Noël Labat. The student visited all the sites in 2022 and found only one remaining plant, which is not stored in any genebank. So WorldVeg partners put a little fence around it – but it was recently lost in a fire. 'Fortunately, they got some seeds and are now producing seedlings for seed multiplication,' says van Zonneveld. 'It is species rescue in live action.'

Veggies in the bank

As head of genetic resources, van Zonneveld strives to ensure the genebank operates professionally in accordance with International Genebank Standards for Plant Genetic Resources for Food and Agriculture. These voluntary standards include maintaining high quality, viable seeds, formalised procedures for characterising germplasm and back-up storage in other genebanks. WorldVeg also partners with Genesys, an online resource for finding germplasm accessions stored in genebanks globally.

Currently, the WorldVeg genebanks store more than 65 000 seed accessions, which includes germplasm of more than 370 species from 155 countries. Twelve thousand of those are indigenous vegetables. However, vegetables still only comprise an estimated 5–10 per cent of germplasm saved in public genebanks – a travesty, given this food group's vital importance for providing life-essential nutrients and gut-healthy fibre.

Reade and I visit the genebank during the anniversary celebrations. Before entering, we take off our shoes and don slippers provided at the door to prevent pathogens entering the facility. The refrigerated storage rooms are a shivering contrast to the warm humid air outside. In a room containing rows and rows of carefully numbered vaults, our guide dips into one of many neatly stacked boxes and shows us a seed packet, with its unique identification number, year of registration and a QR code.

The seeds stored in the genebank are 'orthodox' seeds that can be dried and frozen ('recalcitrant' seeds such as avocado can't or they would die). All are kept under 5 degrees Celsius for mid-term storage (for seed distribution in the WorldVeg genebank) and -18 degrees Celsius for long-term storage (for conservation), with relative humidity levels carefully monitored at around 15 per cent.

'And then dry, dry, dry up to 4–7 per cent seed moisture content," says van Zonneveld. 'That is that most important part of seed conservation.'

Even then, seeds differ in their storage longevity and germination requirements, and seed viability is actively monitored over time. Crops such as cucurbits have relatively short lifespans and could be kept in short-term storage for 10–20 years, or two to three times as long in long-term storage. Others such as mungbean might keep for 50–100 years in mid- to long-term storage. For germination, some crops such as bitter gourd need special heat treatment while others, such as pumpkin, don't need anything other than water to get started. Dr Chutchamas (Chat) Kanchana-Udomkan, director of the Tropical Vegetable Research Centre in Thailand, which works closely with WorldVeg, specialises in papaya. She explains that before drying papaya seeds you need to remove the jelly-like coating, because it contains chemicals that prevent germination.

Type, geno- and pheno-

Saving seeds and keeping them alive is just one consideration. To unlock the genebank's vast treasures, modern genotyping and phenotyping technologies are key. This essentially delivers the means to fast-track centuries-old agricultural practices of collecting seeds from high-performing plants and replanting them next season. Genetic sequencing provides a window into traits embedded in the seed's DNA, and rapid advances have enabled the genes to be identified quickly and cheaply.

'We can genotype anything,' says Shanice Van Haeften, a PhD student researching mungbean crops in Australia, who did an internship at WorldVeg. But while the centre has a vast resource of genetic information, it needs data on the physiological traits expressed by a seed, such as yield, pest resistance or abiotic tolerance – a cumbersome task.

'We have a phenotyping bottleneck,' says Van Haeften. 'We're struggling to get enough physiological information we need to be able to link it to our genetics.'

To help pull the stopper out of the bottleneck, Van Haeften has been experimenting with UAVs (unmanned aerial vehicles, or drones) with specialised sensors to capture data more quickly at a large scale and across the season. Manually, it took her more than nine months to collect data on mungbean biomass – how much light a plant can absorb for production and yield – by harvesting 60 plots, dissecting the leaves and stems and weighing and drying them. Her drone captured the same data, using images derived from different reflective wavelengths, in 25 minutes across 800 plots.

WorldVeg has access to even more advanced phenotyping technology: an automated laser scanner that moves over a field and captures 3D models of the plants under any conditions, generating data that enables diverse traits to be measured. This remarkable piece of kit is made by the Dutch biotech company Phenospex.

'It scans the plants in the field every day, three times a day, and shoots light on the plant,' explains Dr Derek Barchenger, a US pepper breeding specialist. '[It] then measures the wavelengths of light that are returned so we can figure out how healthy a plant is, how much biomass it has, how big it is, how quickly it's growing – without ever entering the field.'

With this system, large numbers of plants can be monitored and then challenged to find the superstars.

'I can see how this plant is feeling in this environment in real time,' says WorldVeg's Schafleitner. 'I can grow it in a hot and humid summer, I can flood it, I can put other stresses on it, whatever, and then select the plants with the most interesting responses.' By doing this, his team has generated improved crops such as flood-tolerant okra and peppers tolerant to heat stress. It's impossible to overstate the significance of this. Attributes such as heat tolerance derive from a combination of traits, not just one. For the peppers, researchers combined 75 different traits ranging from plant height, biomass and pollen variability to leaf area, angle, colour and yield, measured during the entire plant life cycle to select heat tolerant plants.

'You can imagine, without a machine like [the] Phenospex [one], we would rely on only a few traits for selection and probably miss important features,' Schafleitner explains. Collecting 75 traits from plants in the field would require 400 people to work 24/7, he adds. 'Now, measuring plant parameters is automatised and we just need a few gardeners to take care of the plants and harvest the product at the end.'

The sheer amount of information gathered is mind-boggling, with millions of data points. But artificial intelligence and machine learning have unlocked new ways to combine genotypic and phenotypic data to process multitudes of traits simultaneously and provide a more complete picture of how plants respond to stresses. This improves the accuracy of selecting better-performing individuals to enhance the gene pool and help breeders select their desired plants.

For the global good

Ultimately, the goal is to share seeds.

'The genebank is not a museum,' says Schafleitner. 'If we had not shared germplasm in our history, potato would only be eaten around Lake Titicaca and in Europe you would starve.'

WorldVeg has shared more than 700 000 seed samples with researchers and breeders in more than 200 countries, and in 2022 was the largest contributor to the Svalbard Global Seed Vault, on the island of Spitsbergen in Norway. It also provides free seed kits to farmers.

An important advantage of WorldVeg seeds is that they are open pollinated rather than hybridised. 'So you can use the next crop from the last crop, and it fits in with the farmers' culture of saving seeds,' says Rugalema.

While open pollination is unrestricted and creates greater genetic variation, hybrids are controlled combinations of pure lines. They might produce improved plants, but when their seeds are planted, they don't carry the original traits of the parents. So you might get one line that is productive but not adapted to the environment, and vice versa. 'It's like crossing a horse with a donkey,' says Rugalema, 'and then you get this stupid mule.'

In Africa, N'Danikou distributes seed kits to farmers, focusing on traditional vegetables that are not commercially available, and trains them to grow and save the seeds for replanting. To date, he estimates they have distributed more than half a million seeds across ten countries in Africa alone. His personal favourite, yantoto

(*Launaea taraxacifolia*), also known as wild lettuce, is of course one of the traditional vegetables he is dedicated to keeping alive.

'In my position today, I try to revive some of those stories by going back to such communities, and encouraging them to preserve and share seeds with the genebank so we can conserve those treasures.'

✱   *Dog people: How pets remind us who we really are*, p. **40**
Surviving in one place, p. **237**

BORN TO RULER?

Petra Stock

Long-running television show *8 Out of 10 Cats Does Countdown* is an edgy spoof of a letters and numbers quiz.

Hosted by comedian Jimmy Carr, a rotating panel of comics compete, with maths providing the axis for numerous jokes and jibes.

In one episode, resident maths whiz Rachel Riley neatly solves a puzzle, organising a set of six numbers and operators ($+$, $-$, $\times$ and $\div$) to make them equal 576. One of the comedians responds – to laughter and applause from the audience – 'What happened to you? How did you become like this? How many friends have you got?'

According to mathematician–musician Alexander Hanysz, the show reflects wider attitudes and public perceptions about maths. The contestants are consistently 'quite good at the words, they're proud of it and they're creative,' he says. 'And then you get to the numbers, and people revel in being bad at it. I wish we could change this about the world.'

'I'm just bad at maths' has become the popular refrain, a self-fulfilling prophecy that people are naturally predisposed to words or numbers, but never both. Perhaps it's no surprise, then, that the public increasingly doesn't want to do the maths, a fact borne out by statistics showing declining maths participation and performance in Australian children.

Unlike literacy skills, which are widely considered essential, the numeracy realm is regularly dismissed as an inborn ability (the idea of a 'maths brain' hardwired from birth): something that's abstract, solitary and square, with limited career opportunities outside banking and teaching.

In fact, maths graduates are highly employable in a wide range of jobs, and their backgrounds and childhood skills are just as diverse. Is a lot of what we think about the field and its participants based on faulty calculations?

Maths: innate or learnt?

Today, Hanysz is employed at that maths-positive workplace the Australian Bureau of Statistics (ABS), the agency that casts its role as telling 'the real story of Australia, its economy and its people by bringing life and meaning to numbers'.

After a first PhD attempt in his twenties 'crashed and burned', Hanysz spent more than a decade working as a full-time classical pianist, before returning to university for a second go.

From the abstract land of a pure maths PhD, Hanysz eventually found work in the practical world of the ABS. His day job involves sourcing independent, reliable data for telling Australia's story through numbers. An example is an experimental indicator of household spending, drawing on aggregated and de-identified bank transaction data.

It might sound dry, but when Hanysz talks about maths, he leans into the lyrical. 'It's simultaneously an art, a science, a game and philosophy', bundled together with 3000 years of history, he explains. Nonetheless, in most social situations, Hanysz avoids talking about the subject he loves. It makes people uncomfortable, he says, so he talks about music instead.

'I find if you talk about maths, people are scared of it. You get the stories about how people were so bad at maths at high school. Or they think if you're good at maths, you're some kind of freak.'

Hanysz describes his pathway into maths as a 'succession of lucky accidents'. And contrary to the popular misconception that maths ability is innate, many mathematicians recall challenges along the way.

Professor of Mathematics at the University of Tasmania, Barbara Holland, always liked maths, and grew up with the benefit of 'two geeky parents who were both high school teachers', but she says things weren't always smooth sailing.

When she was 14, Holland remembers getting viral pneumonia and missing a month of school. She returned to class just in time to face an end-of-unit test in trigonometry.

'And I'm like: what's trigonometry?' she says. 'Why is everyone chanting 'SohCahToa' like they've joined some weird cult or something?'

Holland did miserably in the assessment, and found herself completely floundering. But thanks to the support of another teacher who was willing to help struggling students, she caught up.

Two lessons from that formative experience have stayed with her. The first is the way in which one good teacher can make a difference in someone's life. The second, that there need to be easier ways for students to catch up.

'Maths is so linear when you learn it in school,' Holland says. 'If you miss some essential building block then it makes everything else quite difficult.'

It's a sentiment shared by Dr Alexandra Hogan, a mathematical epidemiologist at the University of New South Wales. Hogan has spent the last few years writing equations to model the spread of Covid-19: what proportion of the population is susceptible, infectious or recovered, and how limited vaccine supplies might be fairly rolled out for maximum public health benefit. During the pandemic, she worked as part of a large team using science and maths to inform policy, including providing evidence to bodies like the World Health Organization to use in their vaccine planning.

She says that like many people, she found maths at school 'really hard'. But she ultimately persevered thanks to the help of supportive teachers and university lecturers. 'It wasn't an easy thing to do. It was really difficult. But I stuck at it, because I still enjoyed it. And I realised that it's okay to find things hard and you don't have to succeed every time.'

Luckily, persevering with maths pays off, sometimes in unexpected ways.

More than numbers

Surveys of parents, educators and career advisors by the Australian government's Youth in STEM (science, technology, engineering, mathematics) study suggest limited awareness of the potential career paths available from studying subjects like maths or science.

When 730 teachers and career advisors were invited to list as many STEM-related careers they could think of, the top response was teacher or educator (14 per cent). When parents were asked a similar question, the highest response was engineer (11 per cent).

Growing up in South Africa, with a father who was a Hungarian refugee, Dr Éva Plagányi saw maths as a universal language. She always enjoyed maths at school, yet couldn't picture it as a career. 'I had this idea that mathematicians would end up sitting in a little room working on problems all day. I didn't want that, because I was quite an outdoors person,' she says.

She remembers one teacher saying, 'You're going to have a job in mathematics one day ... you're going to be a bank teller.' When she took this information home, Plagányi's father told her instead of counting money, perhaps she could be a scientist. At university Plagányi studied zoology, botany and applied mathematics, even though 'I had no idea what I would actually do with the mathematics'.

She remembers seeing books in the library about how the Fibonacci series can describe the petals of flowers. 'I thought: well, that's great, but it's not very exciting.'

In her second year, a lecturer spoke about using maths to model antelope populations to assist with conservation efforts.

Plagányi says: 'I still remember rushing to his office and saying ... "That's what I want to do!"'

Today, she works with the CSIRO using maths and biology to solve complex, real-world problems about resource use and conservation. Much of her day job involves managing fisheries, combining data collected from the field with equations describing how fish populations change in the ocean. There's no single right answer, but her models can be used to test options and guide decisions on management strategies like fishing quotas, which try to balance the needs of people and the environment.

'For example, if you're harvesting krill, you want to leave enough to ensure that whale populations are healthy,' she says.

A large part of Plagányi's role involves communicating with others, explaining the models to policymakers and stakeholders like Traditional Owners or fishers who are directly impacted by the results of her work. She emphasises that's why it's so important to ensure the models are meaningful, and to have confidence in their results.

'If you get the numbers wrong, you could close a fishery. That's people's livelihood,' she says.

Plagányi's job weaves together her loves of maths, nature and the ocean, and she gets to travel, work with people and spend plenty of time outdoors. But for a long time, she couldn't see anybody using maths in that way: unlike a doctor, lawyer or the ever-popular marine biologist, few can picture a career in maths beyond the limited options of bank teller or maths teacher.

When Kate Simms was at school, plenty of older people warned her against pursuing further study in maths, even though it was her favourite subject. Happily, for Simms, she ignored their advice. She now works on modelling the impact of cervical cancer prevention strategies like vaccination and screening at The Daffodil Centre, a joint venture between the University of Sydney and the Cancer Council – and she still loves the adrenaline rush of solving an interesting problem.

Simms says her university maths cohort have all ended up in diverse careers – everything from cryptography to modelling internet speeds, from public service to finance.

'The skill set and the way of thinking is so flexible, useful in so many different ways ... Even though there's not necessarily a defined job, maybe one of the reasons for that is because there's so many jobs,' she says.

Professor Tim Marchant from the University of Melbourne agrees. The director of the Australian Mathematical Sciences Institute says students – and especially their parents and career advisers – need to know that even though there might not be many jobs labelled 'mathematician', the market for maths graduates is absolutely booming, with the starting salary for many roles at around $100000 a year.

'Jobs involving data science, financial mathematics – there's not enough maths graduates to fill those jobs,' he says.

Marchant lists sectors where strong quantitative skills are in high demand: big technology companies, economics and finance, medical fields, engineering. So much of the modern world is now governed by data that numeracy skills are in high demand.

And it's not only quantitative skills employers are after. Recently, Holland noticed that many of her maths PhD students were ending up in government and social services jobs.

'I was rung up a few times to give a reference,' she says. After a sequence of successful candidates, Holland couldn't help but ask the person on the end of the line: 'This isn't necessarily the sort of job you'd think a maths graduate would get. What attracts you to maths PhD students in these sorts of roles?'

'Well, they're so resilient to being stuck,' the recruiter replied.

Holland reflects that's probably because 'a lot of the time in maths being stuck is the natural state'. 'To try and creatively think your way out of a problem and see it from a different angle is often what helps you make progress.'

Gains and losses

Mathematicians might be applying their craft to solving problems from evolution to economics, public health to environmental management. But what about the rest of us? Do we really need maths?

The ready acceptance of being 'just bad at maths' is under-pinned by a pervasive view that we can get by in life without it. But it's not only the mathematicians, engineers or data analysts who need numeracy skills.

Video game designers and programmers use maths to present everything that appears on screens, from the tiniest pixel to vast virtual worlds. Craft brewers measure water, hops and barley to achieve the perfect flavour profile, and use calculations to estimate heat and alcohol content. Social media influencers analyse data to measure engagement and maximise reach. Fashion designers and carpenters rely on geometry.

Name a task and it probably involves maths.

In fact, research suggests people with poor numeracy skills, particularly women, experience more disadvantages in life.

A large-scale longitudinal study in the UK by researchers Samantha Parsons and John Bynner followed a sample of people born in 1958 and 1970 – about 17 000 in each cohort – through to adult life. By the time people entered their 30s, those with poor numeracy skills were more likely to experience depression and have difficulty finding and maintaining employment compared to those considered to have competent maths ability. They also found a correlation between women with poor numeracy skills and substantial socioeconomic disadvantage – in terms of employment, physical health and a sense of control over their lives – regardless of their levels of literacy.

Parsons and Bynner conclude: 'Poor numeracy skills make it difficult to function effectively in all areas of modern life, particularly for women.'

Maths is fundamental to our everyday existence. A world without maths? There'd be no time, no baking, no page numbers in books, no sports scores, no travel or navigation. No understanding of atoms and density, no rockets, no bridges.

No symmetry. No poetry. No music.

Beyond the practical and pragmatic, maths can offer a measured perspective on the world. Hogan says maths training helps you think about things in a quantitative way, assessing evidence and making objective decisions. 'That's an incredibly useful skill to have.'

Plagányi likes the way maths reduces complexity in the world. It's a structured, consistent way of explaining everything in nature.

Hanysz asks: can you imagine a school system where people had never heard of Shakespeare? And no one was ever encouraged to go for a run, or jump in a swimming pool?

'It's part of our culture,' he says. 'The ability to do maths is part of what makes us human.'

✱ *Predicting the future*, p. **122**
Doing drugs differently: For health, not profit, p. **162**

SURVIVING IN ONE PLACE

Viki Cramer

Trees survive in one place like no other living thing. A tree is, theoretically, immortal. Unlike our own cells, the plant cells that divide to create the thickening growth of trunks and branches are not genetically programmed to die. These cells lie in a thin band, called the vascular cambium, inside each stem, and produce the plant tissues that we recognise as wood. On the inside of the band, the cells divide and differentiate into the tissues that carry water and nutrients from the roots through the trunk to the branches and leaves. On the outside of the band, the cells create the tissues that carry the sugars produced by photosynthesis in the leaves down to the roots. Growth may slow as the tree ages but the vascular cambium can produce new cells more or less indefinitely. This means that a tree does not die: it is killed. Perhaps by a disease that causes its roots to wither in wet soil, or from an outbreak of insects that damage each and every leaf. Maybe years of drought mean that its roots can no longer draw enough water from the soil to keep its leaves alive and they drop, one by one, padding a thick bed of kindling on the forest floor. Then a fire comes along and cleans up the fallen leaves, and the tree along with it.

Rooted in one place, a tree cannot escape the circumstances in which it finds itself. It must endure the cards it was dealt when it fell to the ground as a seed, perhaps not much larger than a speck of dirt on the tip of your finger, and found a safe space to germinate among the grains of soil and bush detritus. But the wisp of seedling that emerges has more than just dumb luck on its side. It has inherited a legacy from its ancestors, carried as a memory within its DNA, of their adaptations to the character of that landscape – the paucity of nutrients in the soil, the highs and lows of temperature, the way rainfall varied between seasons and across decades, how often fires burnt and how hot they ran. It is a legacy that equips the seedling with the physical traits that will allow it to survive in that place.

The transmission on the Hilux begins to whine and I grab the handle above the window to my left as my right shoulder slams into botanist Phil Ladd. My left shoulder rebounds into the upholstery padding around the window's frame as my notepad and pencil launch off my lap. I clutch at my phone to save it following in their path while stretching to grab the pencil before it rolls under the passenger seat in front of me. There are five of us in the cabin, and our collective weight forces the tyres to bite deeper and deeper with each rotation into the titanium-white sands of the Bassendean dunes on the Swan Coastal Plain to the north of Perth. In the driver and passenger seats, fire ecologist Joe Fontaine and his research manager Billi Veber are relaxed: they've been driving tracks like these for the past six years from Jurien Bay, a fishing and holiday village 220 kilometres north of Perth, to Mandurah, a coastal city 70 kilometres to Perth's south. Also in the straining four-wheel drive is ecologist Neal Enright, Joe's first boss after he arrived from the United States in 2009 to take up a research position at Murdoch University. Phil and Neal are now retired. This is meant to be an easy day out in the field.

We are heading east along Clover 'Road', an overstatement of designation for a track that is just wider than the vehicle. We arrive at a crossroads, pull off slightly to park, and Joe and Billi start gathering equipment together from the tray at the back of the ute. Billi is wearing a chocolate-brown contraption of pockets, satchels and brass rings that belts around her waist and buckles around each thigh. It looks almost medieval, like some sort of chastity belt with storage. It was made to be worn by women attending music festivals, she tells me, but it has the perfect combination of pockets to allow her easy access to a tablet, phone, field notebook, GPS and all the other equipment she needs for the day. It also means she can wear snug exercise leggings and calf-high leather boots to deter the pepper ticks, the size and colour of their name, from crawling up between her clothes and skin. Banksia woodlands are renowned for these tiny blood-suckers. Their bites, unfelt at first, rise red and itch for weeks. Phil and Neal tuck the legs of their field pants into their socks then spray their pants and boots thoroughly with insect repellent. I rue deciding to wear pants with such flared hems – there is too much fabric to comfortably stuff into my thin socks. I too spray my calves, ankles and boots liberally. A week later I will stand in front of the bathroom mirror and dab pawpaw balm on the twenty-one tick bites spread over the soft reaches of my belly, upper thighs and inner arms.

Banksia woodland is the kind of vegetation that, to an untrained eye, might kindly be called 'bush' or dismissed as 'scrub'. Its low grey-green shrubs bear small, spiky leaves that scratch your legs as you weave in between them. It's the kind of ecosystem where you must seek delight in small things: the bumpy grey curves of the banksia trunks; the feel of the cone of a firewood banksia under your fingertips, its small mounds of flesh pressing through an intricate black lacing like a fishnet stocking; the anemone hands of a climbing sundew, each fingerling tipped with a globule of sugary glue like a burst of sunlight. And then there are the orchids, some with the legs of a spider, others the ears of a donkey, the pink skirt of a fairy or the hard purple sheen of fresh enamel paint.

There's a buzz overhead, a light plane wafting in repeated circles – perhaps a trainee pilot from the airfield nearby. I look up and wonder if the pilot is looking down and wondering at us: what are five adults doing standing around in the bush out here? To our west are the remnants of vast pine plantations, now being removed and restored to the original banksia woodlands. They have a reputation as a place where cars are dumped and other dirty deeds are done. We are standing near where, in 2015, a guy trying to burn a stolen car started a bushfire. His act of vandalism helped create a perfect opportunistic field experiment: each corner of the crossroads has a different fire history, which is why Joe and Billi have chosen this area to place a series of field plots. It's the kind of serendipity that fire scientists need to make the most of. Conducting their own experimental burns is difficult: fraught with planning, the expense of logistical support from the fire services, and the small windows of benevolent weather needed to light and safely control a fire. And a burn lit for research purposes will behave differently than a wildfire.

Joe explains the day ahead to me. They will measure the diameter of the trunk and the height of each tree within a series of plots, and then compare their measurements with those collected from the same trees six years ago. This will allow them to calculate how much each tree grows in a year, so that they can then estimate their ages. Banksias do not lay down annual growth rings in a way that is interpretable. Even simply measuring the diameter of their trunk at a standardised height is problematic. The banksias branch low, they bend and bulge, each with its own unique figure. 'They're not North American conifers,' says Joe.

Most of the trees here are slender banksia (*Banksia attenuata*) and firewood banksia (*Banksia menziesii*) and, on these soils that are little more than beach sand, they grow to 6 or 7 metres in height. Further north, where summer temperatures are hotter and it rains less, they cannot muster the resources they need from the soil to make it to this height, and they grow as shrubs only a couple of metres tall.

It was in these low kwongan shrublands that Joe and Neal observed a phenomenon that they came to call 'interval squeeze'. Banksias are not the only species that are stunted by the harsh conditions on the sandplains. Few plants in the kwongan reach the height of a human adult, and so anything but the coolest of fires consumes their canopy. Some species, like the slender and firewood banksias, resprout from a lignotuber after a canopy fire. Other species such as Hooker's banksia (*Banksia hookeriana*) cannot resprout. Instead, they are killed by hot fires and a new generation of plants must germinate from seed. It takes five years for a Hooker's banksia to produce its first seeds. As with all banksias, their seeds do not drop to the ground to be stored in the soil, but are sealed securely in the capsules of its woody cones. The hot fires that kill Hooker's banksia burst open the seed capsules on their cones, and the seeds fall to the ground ready to germinate in the ashes of their parents. A new population emerges, all the same age; each plant slowly building its stores of seed.

Here's the problem for Hooker's banksia and other fire-sensitive species like it: to survive in an ecosystem they must have enough time between fires to produce viable seeds. With each year, more cones are held within the canopy, and thus the number of seeds that are stored increases. A fire burning every fifteen to twenty years is the sweet spot for Hooker's banksia, the 'safe operating space' that allows a new generation of plants enough time to produce sufficient seeds to ensure the survival of the generation that follows. How many seeds a shrub produces each year, and how many seedlings survive their first summer, is linked to how much rain falls in the previous winter and spring. As the climate dries and warms, there is less water available in the soil. In response, plants grow more slowly and produce fewer flowers and seeds. It is the combination of fire and chronic drought that really squeezes fire-sensitive species.

Rainfall on the northern sandplains has declined by 20 per cent since the mid-1960s, but in that time the number of seeds stored by each generation of Hooker's banksia has declined by more than half. In 1987, a mature Hooker's banksia would have had around 370 seeds stored in its cones. By 2012, each plant was lucky to have twenty-four seeds ready to fall to the ground in the next fire. For Hooker's banksias, two hot fires burning less than ten years apart mean that they simply do not have enough time to reach reproductive maturity. As their seed numbers dwindle, fire after fire, Hooker's banksia, and species like it, will disappear from the kwongan of the northern sandplains. The more resilient resprouters, for the time being at least, will take their place.

That one stunted species of banksia shrub might be swapped out for another on the remote sandplains north of Perth may seem of little consequence. But the tallest flowering plants in the world, the mountain ash (*Eucalyptus regnans*) of Victoria and Tasmania, rely on the same strategy as Hooker's banksia. As do the alpine ash (*Eucalyptus delegatensis*) forests of the Australian Alps. Both of these eucalypt species are killed by hot fires, with a new generation growing up, all the same age, from seed released from the tree canopy. Mountain ash and alpine ash both need about twenty years of growth before they reach reproductive maturity. But at this age, even though 20 metres tall, they are just at puberty. To have a chance to even begin to reach the mature heights that inspire reverence, to become an ash *forest*, trees need to grow for seventy-five years or more before a fire runs through their crowns. In some places, intense, end-of-summer fires may only happen every two or three centuries. Yet, in Victoria, hundreds of thousands of hectares of ash forest have burnt two or three times in the past twenty-five years. And largely for the same reasons as the kwongan north of Perth: as the climate warms, air temperatures during drought periods are hotter, the fire season is longer and extreme fire weather more frequent. Regenerating stands of ash are particularly fire-prone, the young trees a dense pincushion of leafy, whip-thin stems that are highly flammable. Like Hooker's banksia, a second fire too soon means that the young trees will not have had enough time to flower and develop a seed store in their canopies, and so the prospects for the next generation of trees are dim. The tallest flowering plants in the world will be squeezed out of the landscape, their place taken by more fire-tolerant species: at first acacia shrubs, and then perhaps resprouting eucalypts.

In 2013, a wildfire burnt over 15 000 hectares of alpine ash forest in Victoria, most of which was inside the Alpine National Park. For many ash forests within the park, it was the second or third time they had been burnt within the space of ten years. Forest ecologists and the park's managers held such grave fears for the future of alpine ash that they made the bold decision to use helicopters to spread ash seed over 1857 hectares of the park. Seeding from helicopters is the usual practice in Victorian ash forests after timber harvesting, when the bark and branches and other wreckage of the forest left behind are burnt with a hot fire to create a seed bed. But this kind of intervention had not been used in a conservation reserve before, and seeding had not been attempted at this scale. It was a decision that had to be made quickly, plant ecologist Lynda Prior told me in 2018. The seeds needed to be spread before the first winter after the fire, as they need a snap of cold to prompt them to germinate. The following winter would be too late, because the tiny seedlings would have no chance among the grasses and shrubs that had already rooted their claim on the soil. Here is one of the great ironies of the eucalypts: for trees that are around us everywhere, eucalypts are not very good at getting anywhere. It's not only apples that don't fall far from the tree – eucalypt seeds are simply released from their capsules and fall to the ground. Even the lightest seeds dropped from a height of 40 metres only travel 50 metres or so. A study that assessed how far eucalypt species are able to migrate across the landscape found that they could move, at best, 2 metres each year. Even though there were large areas of the national park where mature alpine ash survived, there was little chance that their seeds would ever arrive in sufficient numbers in the burnt areas to ensure the recovery of the forest.

There was another hesitation: the seed had been collected from outside of the national park, and may not carry the best genetic adaptations – that memory in their DNA – to the environmental conditions within it. Best practice restoration is to use seed collected locally (although this is changing now that climate change is shifting the boundaries of temperature and rainfall). It was a confronting decision, said Lynda, but the overwhelming motivation was simply to get the forest back. Seed sourcing was coordinated across the state and matched by altitudinal level. The project was successful in re-establishing young trees in the national park, despite limited seed supplies meaning that the helicopters could deliver less than 1 per cent of the amount of seed that would have fallen from mature trees over the same area. But there is a caveat: this kind of one-off intervention has no chance of becoming the norm over large parts of the landscape. Apart from the enormous costs involved, there are simply not enough mature trees remaining from which to collect sufficient amounts of seed.

There's that squeeze again. Even in the coolest forests on the continent, it's becoming too hot and too dry. The big eucalypts are growing more slowly, and the young eucalypts get burnt before they are mature enough to produce seed. Even if we, as a society, decided to devote the money, time and energy necessary to restore these forests after fires, we simply do not have the ecological resources – enough seeds – to do so. These giant trees, our megaflora, managed to survive the massive changes in climate and fire that saw the extinction of the Australian megafauna – the diprotodons and kangaroos 2.5 metres tall and crocodiles 7 metres long – at the end of the Pleistocene. Yet it is starting to look as if we may witness their twilight, a blinking kind of existence where trees like alpine ash can only survive amid the refuge of the deepest, coolest corners of the landscape.

In the roulette wheel of evolution, it would seem that the resprouting eucalypts made the safer bet when they laid their buds down deep under their bark or nestled them safe in a woody clump of lignotuber. But the resprouting eucalypts that would be expected to eventually fill the footprint of a collapsed ash forest – species such as messmate stringybark (*Eucalyptus obliqua*), mountain gum (*Eucalyptus dalrympleana*) and brown barrel (*Eucalyptus fastigata*) at wetter sites, and snow gum (*Eucalyptus pauciflora*) and broad-leaved peppermint (*Eucalyptus dives*) at drier sites – are already challenged by the frequency of fires at one degree of warming. The resprouters are being squeezed too.

A study of the drier and shorter eucalypt forests common in eastern Victoria found that two severe wildfires less than a decade apart began to erode the ability of resprouting species to recover. After the second fire, many trees showed signs of a sort of 'resprouting exhaustion syndrome', especially the medium-sized trees that sprouted from both their lignotuber and epicormic buds after the fires. With their reserves depleted by resprouting after the first fire, they simply did not have the energy to send out new growth after the second. The first fire triggered a mass germination of seedlings that was not repeated after the second. When hot fires come close together, even the resprouters don't have enough time to burst buds, grow leaves and flowers, then harden their gum nuts and release their seeds.

I ask Tom Fairman, the forest scientist who led the Victorian study, how these dry forests of resprouting eucalypts will fare in the future under climate change. It's a question the research community is now beginning to grapple with, he responded. If the question is as simple as 'Are these trees resprouting?', then, yes, they are. Forests of resprouting eucalypts won't be wiped from the landscape the way that fire-sensitive ash forests will if they are burnt frequently. But the idea that resprouting eucalypts are 'fire weeds' and that they will be fine, even as fire seasons become longer and fires burn hotter through drier forests, is unlikely to hold. If small and medium-sized trees never get a chance to grow tall enough so that their canopies are not killed by every fire that flickers up their trunks, they might still live, but they will never grow to their full potential.

Tom tells me about one of his research sites, a rocky, dry, north-facing slope that is a tough place for eucalypts to grow. It had been 'really hammered' in the two fires, and he was uncertain whether to include it in his analyses. Black stumps were sticking up through the chest-height foliage, with dense shoots sprouting 'like crazy' from big lumps of lignotuber. The size of the stumps and their root mass suggested that, before the Victorian fires, there had been tall trees here. Yet what was growing now on the slope looked like a eucalypt shrubland, not a eucalypt forest. Tom searched for flowers and fruits and discovered that he was walking through a patch of the same stringybark species (red stringybark, *Eucalyptus macrorhyncha*) he had measured at his other sites, but just growing in mallee form. He kept the data in his analyses. For him it signalled the trajectory that these forests will follow as the climate becomes drier, the trees become more stressed, and they are burnt by fires that are more frequent and severe. This, he thought, is what the forest might look like when it gets as bad as it gets. When it loses hold of our idea of a tree: a single stem, tall, of having to look up to see into its canopy.

When Tom presents his research at ecological conferences he is cautious not to voice a value judgment about these changes. Ecologists are reticent to label one form of an ecosystem as more important, more worthy of our attention and admiration, than another. Ecosystems are not static in time – there is no 'balance of nature'. The species within an ecosystem are continuously reorganising after disturbance and adapting to changes in the environment around them. A patch of stringybark is still a patch of stringybark even when it grows as a multitude of stunted stems.

Then Tom says what I am thinking: but it's not a forest.

Billi is calling out a stream of directions and questions across the banksia woodland plots: 'It's a bank att. Live height? Dead height? Mort 15 per cent.' She is standing in the centre of the plot and directing the men to find and measure a specific *Banksia attenuata*, 'bank att' for short. Phil sprays a purple dot of paint to mark a permanent spot on the trunk. Neal then wraps the tape measure tightly around the place marked, and calls back the measurement of the tree's girth to Billi. Joe stands beside Billi and lifts a bright yellow laser rangefinder to his right eye to measure the height of the tree. The team is interested in the height of both the living and dead branches. Mort is short for mortality, an estimate of the percentage of the tree's branches that are dead. It is fiddly, repetitive work. It's mid-April and, now that the sun is high, the day is hot. This type of woodland is what ecologists call *open*, with trees standing apart, and even when I stand up close under one of the banksias the shade is sketchy.

There's a sheen of sunburn cream and sweat on Joe's forearms as he walks over to me standing under my banksia, remembering why I began to hate the hot repetition of field work, and my admiration for those who still find purpose and joy in the long, uncomfortable days of marking, measuring and recording.

What they are documenting, explains Joe, is the slow erosion of the banksia woodland. Tree cover is decreasing and the balance between trees and shrubs is changing. The woodland will eventually become a shrubby heathland, like those on the sandplains to the north where Neal and Joe first collaborated over a decade ago. There are the short and sharp events, the winter droughts, summer heatwaves and following fires that bring individual trees to the brink of death. But here, after four decades of declining rainfall, the changes are ongoing and existential: there is just not enough water within the ecosystem to support the growth of trees.

The slender banksias will still be here, I ask Joe, but just in their shrub form? Sure, it won't be a woodland anymore – it will become more like a kwongan shrubland – but will anything be *lost*?

Joe pauses a moment before answering. There are species here that need shade, and once the woodland loses the tree forms of banksia, these shade-needy species will no longer be able to survive. The small wonders of this place, the orchids and the sundews, will disappear. Species too small and short-lived to have a chance to invest in more than one strategy, species that rely on the success of others to ensure their place in the ecosystem is secure.

Losing the big structural elements of an ecosystem is like knocking down a few of the supporting walls of your house: the building might still be standing but, with its integrity gone, it will begin to shift and crack. The plaster might fall, flake by flake, from the ceiling, just a bit at a time, not enough for you to pay it much attention. You could be safe for a while, but you're likely to wake one morning to the sound of the roof collapsing around you.

Ecologists describe this loss of species from an ecosystem, one after another, as a cascade. Stressed banksias produce fewer flowers, meaning less nectar and pollen for honey possums, the sprites of the marsupial world who weigh no more than three cubes of sugar. Stressed banksias also produce fewer seeds, meaning less food for Carnaby's black cockatoos, already endangered. When, in spring, the birds fly back to the woodlands where they nest, the large, old trees that have hollows deep enough for them to safely fledge their chicks are falling down, forced by the weight of the wind or, in agricultural fields, the push of a dozer. Or they succumb to fire. Or they just become too haggard with age and drop their hollow-bearing limbs. A tree needs a long time to develop these kind of hollows – estimates in salmon gum, a favoured nesting tree, range from 150 to 450 years.

Over forty species of vertebrates – mammals such as Leadbeater's possum and the yellow-bellied glider, and birds like the threatened sooty owl – rely on the hollows and cavities that only begin to form in mountain ash when they reach 120 years of age. Critically endangered regent honeyeaters feed and nest in the tallest trees with the largest trunks in the box and ironbark woodlands of Victoria and New South Wales. The honeyeaters are dwindling in number along with the mature red ironbark, yellow box (*Eucalyptus melliodora*), grey box (*Eucalyptus microcarpa* or *Eucalyptus moluccana*) and Blakely's red gum (*Eucalyptus blakelyi*), whose flowers they rely on for nectar.

It's not just for our own sense of place that trees need to grow tall, with a single stem and a canopy branching high above the ground.

✱ *Call of the liar*, p. **198**
   Everything starts with the seed, p. **219**

'GIVE THE ESPRESSO A LITTLE SWIRL': THE VERY PARTICULAR SCIENCE OF A GOOD CUP OF COFFEE

Bianca Nogrady

A good espresso coffee is sexy as hell. It flows out of the machine at a languid pace, initially dark and brooding, before shifting into a golden foam that would bat its eyelashes at you if it had them. Once settled in the glass, it breathes out and releases an intoxicating scent that is earthy and sweet, capturing everything from the aromas of freshly sawn timber and rich dark chocolate to delicate floral and fruity cherry scents.

Making the perfect espresso is almost a high art, especially for those who practise at the elite level – and Australia has plenty of those. Ranked one of the biggest coffee markets in the world, the Australian coffee market is worth more than $9 billion and Australia coffee fiends sip and savour around 2 kilograms of coffee beans each year. One in four of us reckon we can't get through the day without a brew.

The hard science of the drink so many of us love and rely on is simultaneously fascinating and illuminating. Understanding it can help elevate even the cheapest bean and most basic coffee-making equipment into something far greater than the sum of its parts.

To begin at the beginning, with the bean. The *Coffea* plant – *arabica* or *robusta* – is famously picky about its habitat. It wants plenty of sunshine but without the heat, so it only grows in the cooler, high-altitude areas of the 'coffee belt', a band that extends between the tropics of Cancer and Capricorn.

These growing conditions allow for long, slow development of the bean, according to the University of Queensland agricultural scientist Professor Robert Henry. The bean – or the coffee cherry, as it is known at this point – is so precious about its growing conditions that even the microclimate within an individual tree makes a big difference to the quality. Henry and colleagues found that the coffee made from cherries growing at the bottom of an individual coffee bush was much better than the coffee made from those at the top of the plant.

The contrast between the conditions in which the bean is grown and what happens to it after it is picked must come as quite a shock to it.

'Coffee is prepared in a very extreme manner,' says Professor David Hoxley, whose day job as a physicist at the La Trobe University focuses on the world of semiconductors, but who was lured into the coffee den as something of a scientific side hustle. 'It's heated up to the point where the bean explodes, maybe twice, and then it's ground, like brutally.'

First, the cherry is rid of its plump red skin and pulp to reveal the pistachio-coloured bean. This is then fermented, dried and hulled. What's left is a pale golden bean that gives little hint of its inner chemical glory; possibly thousands of chemical compounds that are still being discovered and described.

'Roasting is everything'

Now, the roasting. At temperatures of about 200 degrees Celsius, interesting things happen to the cellular structure of the coffee bean, according to the chemical scientist Dr Monika Fekete from Breville Australia. 'After what we would call first crack in roasting, the cell walls explode with steam and other gases, like popcorn,' she says.

The first flavour that develops in roasting is acidity but then a chemical process called the Maillard reaction happens. This is the interaction between amino acids, sugars and heat that gives us everything from the comforting aroma of toasting bread to the mouth-watering smell of a sirloin on the barbecue. 'As you keep roasting further, you get more chocolatey, nutty, caramelly flavours,' Fekete says.

Roasting is everything in coffee; it's why there are competitions just for coffee roasters. Even a fairly ordinary bean can be elevated by the careful application of heat, although that won't provide the glorious complexity and richness associated with fundamentally good beans. The roasted bean is packaged in airtight and lightproof packing – oxygen, moisture and light are the enemies of good coffee, Fekete says, so keep the coffee beans in their original packaging as long as possible and only take out what you need to grind.

The physics of coffee grinding

Then, it's time to grind.

But don't rush. 'Often coffee is best around 14 days after roasting,' says Melissa Caia, a judge at the World Barista Championships and a coffee expert at Melbourne's William Angliss Institute – a specialist training centre for industries including food and hospitality. In general, coffee beans should be used within two to three weeks of being roasted. They should also only be ground just before brewing to preserve their freshness; Caia says even grinding the beans a few hours before use is too long.

Grinding is another process that gets physicists excited, so much so that a study published last month likened the physics of coffee grinding to what takes place in the ash clouds of volcanoes and on a moon of Saturn.

It's about triboelectrification and fractoelectrification, which describes the build-up of static electricity in coffee as the beans are fractured and the fine particles jostle together. It has relevance for space exploration – the build-up of particles that could cling to cameras on a Mars lander, for example – and for volcanologists, who seek to understand the behaviour of rocks and dust during eruptions.

In the terrestrial coffee world, that build-up of static electricity makes ground coffee clump together in the basket, and 'when you have these clumps that form, you inherently get variable espresso', says a computational materials chemist, Professor Christopher Hendon, from the University of Oregon in the US.

His study tried to vary a lot of things – the bean, the roast, the moisture content, the grind size – to see if that changed the static built up. They found the single biggest factor was moisture: the drier the bean, the more static you get. A simple spray of water onto the beans before grinding would solve the clumping problem.

And here's where the die-hard coffee tragic might roll their eyes and point out that serious coffee-heads have been doing this for years. It even has a name – the Ross droplet technique.

The size of the coffee grinds is the subject of constant experimentation, from the home brewer to the world champion barista. It's incredibly complex, which really lights up the dopamine centres in the brain of Dr Jamie Foster, an applied mathematician at the University of Portsmouth in the UK.

Applied mathematics focuses on developing mathematical models of physical processes with a practical goal in mind. Foster admits that coffee is a tricky topic, because those chemical and physical processes intersect with the messy, subjective sensory sciences.

Grind size is key because it governs how the water flows through the coffee as it is being brewed. That's described by a law that's more than 100 years old, called Darcy's Law, that's more commonly applied to people trying to extract things like oil from sediments – 'Essentially, anywhere you've got a fluid flow in a porous media.'

The general rule is espresso machines need a finer grind than stovetop espresso 'moka pots', which in turn need a finer grind than the French press plunger coffee maker.

Foster's modelling efforts delved deeper into this and identified a sweet spot of grind size, where the coffee particles aren't too big (which means overall less surface area in the coffee is exposed to the brewing water) or too small, where the grounds clog together and reduce the overall coffee particle surface area available.

They found an 'optimised tasty point' (yes, it is actually called that), where all these factors come together to deliver the greatest amount of extracted coffee from a set weight of grounds. Because each bean, grinder and espresso machine is different, it means experimenting with grind size, the amount of coffee and extraction time to find the perfect combination for your bean, machine and flavour preferences.

Foster says the coffee industry is a bit snooty about his scientific approach. 'They kind of felt like, "I don't want your dirty optimum, I want to faff around and find it for myself and create things,"' Foster says.

At last – the brewing

Finally, it's brewing time. Caia says filtered, unboiled water is a must – it removes sediment and unwanted minerals. For espresso machines, the water temperature must be controlled within a narrow range – no cooler than 88 degrees Celsius or hotter than 97 degrees Celsius.

Many people, like my coffee-tragic husband, have a complex ritual of tapping and tamping the grounds in the basket used in an espresso machine. Caia says it's important to ensure even distribution of the ground coffee in the basket, and 'the idea with tamping is just to flatten and level so that you've got an even ground for the water to penetrate through'. But tamping won't fix a bad bean or roast, she warns. It will, however, make sure you get the most flavour from the bean.

The proof of all of these countless steps and processes is in the golden-brown layer of foam on the finished espresso shot: the crema. Chemically, the crema is an emulsion of tiny bubbles of carbon dioxide which, when forced through and out of the grounds, in an espresso machine or stovetop maker, get coated in the proteins and oils of the coffee.

As the espresso lands in the lower-pressure environment of a cup, those bubbles of carbon dioxide rise to the top, much like the fizz when a bottle of soft drink is opened. The fresher the beans, the more carbon dioxide they contain, and the better the crema. The colour doesn't necessarily matter as much – some high-end beans naturally produce a lighter crema – but its persistence and flavour does, Caia says. 'We recommend people to just give the espresso a little swirl so that you can blend the flavours through the coffee.'

What happens next is up to you. Fekete starts the day with a double-shot flat white, then just one filter coffee later on. 'I try to watch my caffeine consumption,' she says.

✱ *Everything starts with the seed*, p. **219**
A chemist's guide to optimism, p. **261**

CHIPS NOW COME IN FLAVOURS LIKE CHEESEBURGER. HOW DO FOOD CHEMISTS GET THE TASTE RIGHT?

Belinda Smith

I'm a real sucker for a packet of chips. Nothing grabs my attention faster than the unmistakable sound of someone busting open a bag of salty, crunchy potato deliciousness.

When it comes to flavour, I'm not too fussed. I'll eat most.

Traditional flavours are most popular in Australia, with older chip eaters favouring original, and salt and vinegar preferred by younger folk. But now, the flavour range is broader – and more adventurous.

No one knows this better than Hamish Thompson, who runs the Museum of Crisps, a website that's so far logged just shy of 1400 different bona fide flavours of chip (or crisp, depending on your preferred term).

'There's brussels sprout, cappuccino, lamington, and whisky and haggis, to name a few of the weirder ones,' Mr Thompson, who lives in Tasmania, says.

In a grocer near the ABC Melbourne office, some of the far-out flavours are full meals, such as bolognese, cheeseburger and beef rendang ... which taste uncannily like bolognese, cheeseburger and beef rendang.

So how are these complex chip flavours made, and how do food chemists get them tasting so close to the real deal? High-tech laboratory equipment does play a role in this process, but there's more human input than you might expect.

A potted history of chip flavours

The earliest known published recipe for the humble potato chip can be attributed to an English eye doctor who had a penchant for cooking.

'Crisps were originally invented by a guy called William Kitchiner – he was kind of like the celebrity chef of his time,' Mr Thompson says.

In 1817, Dr Kitchiner, perhaps a victim of nominative determinism, published a cookbook called *Apicius Redivivus, or The Cook's Oracle*, which became an international bestseller.

Under 'Vegetables', Dr Kitchiner outlines 'Sixteen Ways of dressing Potatoes', including a recipe for 'Potatoes fried in Slices or Shavings'. Once the thinly sliced spud is fried until crisp and excess oil is drained off, he instructs the cook to 'send them up with a very little salt sprinkled over them'.

'So there you have an early reference to ready salted,' Mr Thompson says.

There wasn't much movement on the flavour front until the 1950s, when barbecue – developed in the US – burst onto the scene. They were incredibly popular. People wanted more.

'So then you start to see the emergence of new flavours like salt and vinegar, and then along comes prawn cocktail, and cheese and onion,' Mr Thompson says.

Such classic flavours reigned until the 1990s, when advances in food chemistry meant almost any food could be reduced to its chemical components. As grunge, beanie babies and hypercolour T-shirts swept the globe, so too did new and unusual chip flavours.

'And now,' Mr Thompson adds, 'pretty much anything goes.'

How to write a flavour 'recipe'

Crucial to the relatively recent chip flavour explosion was a laboratory technique called gas chromatography–mass spectrometry (more on that in a sec). And while it's commonly used by food chemists like the CSIRO's Joanna Gambetta to craft flavours, the process still needs people. The first step involves a 'sensory panel' of trained tasters, preferably comprising a broad range of people.

There's huge variability in what we can and can't taste. Some people have genes that make them more sensitive to bitter compounds, for instance, or their cultural background means they eat more sweets, so are less likely to perceive sugar.

As panellists eat the food of interest, they comprehensively note what they taste and smell, because flavour is more than what happens on our tongue, Dr Gambetta says.

'Aromas are important. A lot of what we perceive is what we get through the nose. So we get them to tear a food apart into its different components. Is it sweet? Is it salty? Is it bitter? Does it smell like oranges or red fruits?'

The panel comes up with a 'map' of taste and smells, then the flavour chemists get involved.

'We'll go into the lab and start thinking: "Okay, where can we source these flavours and aromas from?"'

Some are easy enough. Chemists can nail the floral, fruity notes of fresh strawberries with a handful of compounds, each of which contributes fruity, buttery, leafy or caramel aromas. But not all flavours are as straightforward to build. And this is where gas chromatography and mass spectrometry (GC–MS) come in.

GC–MS separates and accurately identifies the hundreds – even thousands – of compounds in a sample. The principles of the technologies have been around for more than a century, and it's now widely available and a staple in analytical chemistry laboratories.

It works by heating a tiny amount of liquidised food to liberate aromatic molecules.

'Imagine you have a hot cup of coffee,' Dr Gambetta says. 'The vapours carry the smell of it.'

The vapourised molecules are pushed through a long coiled pipe where they separate, with each type of molecule travelling at its own speed. At the end of the coil, the molecules hit a detector, which takes that signal and displays it on a computer as a peak on a graph. A taller peak means more of that particular molecule was present.

'And then through software and databases that other people have composed over the years, we can click on a particular peak, and it will give us [the compound's] name and identity,' Dr Gambetta says.

Alongside the machine doing its thing is ... a person. They sniff from a small, plastic cone poking out the chromatograph's side and try to identify the compounds as they're separated. That's because a small peak on a graph might actually smell incredibly strong to a real live human, Dr Gambetta says.

'Our noses are so spectacular that sometimes we are capable of perceiving things so much better than any equipment that we have.'

This is especially true for very pungent molecules that are present in teeny tiny amounts, beyond the capacity of a machine to pick up, but pack a stinky punch to the human nose.

Combining the human experience with laboratory technology output means food chemists can start building the flavour.

Building a flavour from the 'recipe'

So you know what molecules need to go into your food flavour. Surely it's easy from here, right?

For simple flavours, sure – they can be developed relatively easily. There are shortcuts. Smokiness gives the impression of ham or bacon.

What we perceive as a flavour is also affected by what we see. A reddish colour will make a chip taste more meaty. Yellow gives it a cheesiness.

'It's the same with packaging,' Dr Gambetta says.

'If the packaging says ham and cheese, then you will already be conditioned to think it tastes like ham and cheese.'

But trying to make a complex flavour like cheeseburger, Dr Gambetta says, 'takes so much longer because the combination of ingredients that you will have to use to replicate it is much more complicated. Then you also have more potential for things not to be quite right – not to be quite at the right concentration or to be too low or too strong.'

Like normal cooking, it's possible to have too much of a good thing. Dr Gambetta used to work with compounds called thiols. In small concentrations, these molecules smell like passionfruit. Too much, and they start to smell like cat pee.

Molecules can also modulate the taste and aroma of other molecules.

Essentially, really nailing a flavour involves plenty of trial and error, with each iteration put through the ultimate test: does the consumer like it? Once chemists are happy with the recipe, it is standardised and replicated.

It's just like coming up with a new recipe at home, Dr Gambetta says. 'The difference is that we play with way bigger toys than you would have in your kitchen.'

✱ *Rural pharmacy placement*, p. **109**
Are psychedelics a treatment for long Covid? Researchers probing this mystery don't have the answer yet, p. **180**

A CHEMIST'S GUIDE TO OPTIMISM

Ellen Phiddian

I am the sort of fool who bets against rain in the tropics. I decided not to pack a raincoat or an umbrella on this trip to Cairns; now it's day one and I'm faced with crossing town on foot in a downpour.

Fortunately, there's an open supermarket right near my hotel, with a rack of foldable umbrellas next to the door. I buy one for $10. It keeps me and my laptop dry for the next four blocks. It occurs to me as I reach my destination that rocking up to a sustainability conference with a brand-new, mostly plastic umbrella – the third such one I own – is perhaps a worse look than turning up soaking wet. I pull the tag off surreptitiously, and notice that there's a small tear in the umbrella canopy where it had been punched through. So it's also on its way to disposable.

Over the next three days, my umbrella becomes something of an albatross, dangling from my wrist. I'm in Cairns to cover the first Australian Conference on Green and Sustainable Chemistry and Engineering: several hundred people gathering to discuss how chemistry can make the world sustainable.

The crowd is mostly research chemists, but there are also engineers, patent attorneys, industry reps – anyone who works with molecules is welcome.

The fact that it's happening at all is surprising to some. Chemistry is a filthy science. It's a discipline that evolved from alchemists in smoking laboratories, changing their greedy fascination with gold and immortality into a systematic fascination with money and health. And according to some of the attendees, it's now become something of the 'monster under the bed' in environmentalism.

Chemistry's track record is thoroughly mixed. In the past century, chemists have turned air into ammonia, and then fertilisers and chemical weaponry. They've invented molecules that become life-saving medications, which large companies sell at high cost to make huge profit. They've spun carbon atoms into plastics that facilitate every aspect of modern life, and choke the oceans. And they haven't always been as enthusiastic as other scientists about sustainability.

'When Rachel Carson started the whole environmental movement with *Silent Spring*, I think the chemical community did a poor job embracing that process,' Dr John Warner, one of the conference headliners, tells me.

I've met Warner with his long-time collaborator and the other conference headliner, Professor Paul Anastas, in a humid corridor during the 'one half-hour' they both have free. But neither of them act like they're in a hurry. They also keep quoting each other while we speak. They're childhood friends, from the same town in Massachusetts, who have now been working with each other for decades. And at the start of their careers, they watched chemists distance themselves from the environmental movement.

According to Warner, the decades following the 1960s saw chemists double down on their craft, and ignore its effects. Anastas believes that by the early 1990s, 'everybody but the chemist' was trying to solve environmental problems.

'The chemists only had one role in the early '90s, and that was to measure how bad the pollution was,' he says.

For these two, it seemed absurd. So many of the biggest environmental challenges – recycling, carbon dioxide, energy storage, monitoring – fall within the domain of chemistry.

'The people with the most power, the most influence, the most ability to actually change this equation, weren't involved,' says Anastas.

But there were enough interested chemists to start a movement: green chemistry. Warner and Anastas spearheaded its growth in the US, publishing *Green Chemistry* in 1998. In it, they outlined 12 principles for chemists to follow when doing their work. Here's an abridged version of those principles:

1 Prevent waste
2 Be economical with atoms
3 Avoid hazardous processes
4 Design safer chemicals
5 Safer solvents and auxiliaries
6 Design for energy efficiency
7 Use renewable raw materials
8 Avoid unnecessary modification
9 Use catalytic reagents
10 Design for degradation
11 Real-time analysis for pollution prevention
12 Inherently safer chemistry for accident prevention.

Instead of just focusing on the final product, made as cheaply as possible, green chemists begin by considering each of these principles.

It's hard work. But Warner, with over 300 patents to his name, reckons it hasn't slowed him down. In fact, done properly, he believes green chemistry saves money and time. No one throws resources into research and development, only to find their promising new reaction uses a solvent that's slightly too toxic when it hits the manufacturing wing. No one spends millions in litigation or redesigning when people and governments find their practices are polluting waterways.

'Had that first chemist understood green chemistry – not always, things are still going to slip through the cracks – but there is a much better chance that ... they will avoid those profound mistakes that become very costly to industry,' he says.

One of Warner's inventions is a way to dodge hair dyes, many of which are toxic to the waterways they end up in. Warner's treatment, derived from velvet beans and called Hairprint, uses melanin instead. As we age, the melanocytes in our scalps stop making melanin, the pigment that gives our hair its colour. Hairprint returns that melanin – and because it originally sat in our hair, it's incorporated back in the same way, resulting in the same colour one's hair originally was. For now, it only works with eumelanin, which yields black or brown hair – sorry, pheomelanin-stacked redheads and blondes.

Warner was the first person to try his hair treatment. Similarly, he dug up his own driveway to test a new asphalt binder. The non-toxic, environmentally friendly substance undoes the hardening oxidation reaction that asphalt undergoes while it's sitting in our roads – making it soft and easy to reuse. Now sold as Delta S, the binder lets road builders re-use much more of their old asphalt. Neither of these products is perfect. Warner hasn't yet made something that addresses all 12 principles. 'Right now, there are people dying of cancer,' he says. 'If I can get to market stopping this thing from being a carcinogen, and it's still not the most biodegradable, that's the way science works.'

Anastas has spent more of his time with the US Environmental Protection Agency, but there are a few inventions that bear his stamp. One of his PhD students, Stafford Sheehan, has developed a process of particular note: it reacts carbon dioxide, CO_2, with hydrogen, forming a mixture of alcohols, water and a class of chemicals called alkanes. This has now spun out into a business called Air Company, which takes CO_2 captured from industrial plants and turns it into vodka, hand sanitiser and perfume. For now, the goods are small fry – but Air Company plans to make jet fuel. Given jet fuel pumps CO_2 straight back into the atmosphere once it's burned, this would be circular rather than carbon-negative – but it's certainly an improvement on our current, one-way fuel.

The challenge, now, is to get everyone else on this wavelength.

Green chemistry has flourished in the past 25 years, but it's not yet the standard.

'It is a paradigm shift. But I'd like to believe that it's an evolution as opposed to a revolution,' says Warner.

'Evolution's good. Revelation's better,' says Anastas.

Both Anastas and Warner understand that they need to take other chemists with them. It's why, when Professor Colin Raston wrote to them in the late 1990s about starting the Australian green chemistry movement, they got back in touch promptly to lend their expertise. A quarter of a century later, they've given up their 4th of July celebrations to come to a rainy Cairns and speak at the conference Raston is co-chairing. Shut in from the rain, the venue is warm, dark and damp: perfect for fermenting ideas.

A handful of inventions from two chemists – even really brilliant ones – won't stave off environmental destruction. But thousands of inventions, from people all over the world? That's got potential. This is why Warner thinks the most important thing now is to change chemistry education.

'If we change education, there's enough diversity of people that can go out there and start solving problems. And I don't have to prioritise, because we get them all,' he says.

Doing the right thing

After I've finished speaking with Warner and Anastas, I realise my new umbrella is missing. I find it in one of the conference rooms: it had rolled under a chair I was sitting on a few hours earlier.

The person who made its handle didn't care about the person who made its canopy. They don't fit together – it's awkward and loose. The person who attached the price tag to it didn't care that they were puncturing the fabric. None of them cared about me, struggling to keep this bad umbrella stable in a storm that evening. And I didn't care about any of them, or I would have spent more than $10 in the first place, and kept a closer eye on it.

Professor Edward Buckingham, now director of engagement at Monash University's business school, is familiar with these chains of apathy. Buckingham has the air of a polymath – he quotes everyone from Machiavelli to Marx to Manuel (the young man he was speaking to when I approached him for an interview). He started his career as a materials scientist, and was running a manufacturing line in France when he started to change his mind about chemistry.

'We made the tubes and the stoppers that effervescent tablets come in, and we consumed about 18 tonnes of polymer every day: mostly polypropylene, but polyethylene and a little bit of polystyrene as well,' he says. 'I was walking in the forest of Fontainebleau near my home, which is a forest just south of Paris. And I found one of my tubes lying in the leaves. That upset me, because I thought: that's where my product ends up. We don't recycle it.'

He took it to his boss, who dismissed him – saying their company's pollution was insignificant compared to others, and the tubes made people's lives more convenient. It started Buckingham thinking about what a limited role he had, where his primary goal was efficiency.

'There's this trade-off between efficiency and effectiveness,' says Buckingham. Efficiency, he says, quoting management author Peter Drucker's definition, is 'doing things right' – while effectiveness is 'doing the right thing'.

'What happens further down the value chain to the product that I produce? What happens when it combines with other products and creates some unintended consequences? Efficiency doesn't deal with that.'

Buckingham is now an ethnographer, but he still runs workshops on business design for budding scientists – having started in sciences, the field isn't foreign to him. He's at this conference to try and get pure chemists thinking about how industry sees their work, and hear the chemists' perspectives.

One of those chemists is Professor Maria Forsyth, from

Deakin University, who works on one of the most ubiquitous and problematic pieces of chemistry on the planet. Lithium, as the lightest metal, is undeniably the easiest way to make a powerful battery. It's hard to imagine electric vehicles or phones made with anything other than lithium-ion batteries. But in situations where weight isn't as crucial – like a home or grid-scale battery, or even an electric lawnmower – we don't need to use such a scarce resource. Abundant sodium, sitting just below lithium on the periodic table, shares many of its features for a much lower price.

'If you can have a manufacturing plant that can make lithium-ion batteries, you can swap in sodium-ion batteries,' Forsyth says.

Forsyth is rushing to get things done right. In a 40-minute presentation to the conference, she packs in enough information to fill six undergraduate lectures on sodium-ion batteries. She goes even faster to get more words in when I interview her: I need more of the technical stuff explained in detail.

Sodium-ion and lithium-ion batteries share all the same parts, but the chemical make-up of those parts is different. This is why sodium-ion batteries are still, for now, only buyable for specialist applications: chemists haven't optimised them yet. Forsyth believes we're about five years out from full commercialisation – and that the batteries can be optimised in more than just price.

'We have an opportunity to make the devices that we make more sustainable, and more easily recyclable,' she says.

Take the battery anode, which supplies electrons in one half of the battery. In lithium-ion batteries, the anode is typically made of graphite. Sodium doesn't play as nicely with graphite – the atoms are too big – but it does work with another form of carbon, called hard carbon, which can be made from junk: waste biomass. Forsyth and colleagues have been toying with anodes made from biochar, green waste and even unwanted textiles.

Then there's the current collectors, which connect the anodes and cathodes with the outer circuits. Lithium needs both copper and aluminium in its current collectors, but sodium can use just recyclable aluminium. Or electrolytes, which transfer positive ions within the battery. Forsyth and colleagues are designing solid-state electrolytes for sodium batteries, which will be safer and therefore much less energy- and time-intensive to manufacture.

Forsyth wants to see onshore battery manufacturing in Australia – but she says we should 'think about what we're doing' first. Once an industry is established, it's much harder to change it to be sustainable. 'One of the things that drives me at the moment is making sure that what we create now doesn't create a problem for the next generation,' she says.

For Buckingham, paradigm shift is about perspective – and getting other people to see things from your line of sight.

'It's being able to reposition yourself in order to see a pathway through,' he says. 'That's what a paradigm shift is all about. Now, very often, you sail off down that street on your own, [and] no one else follows.'

So it's crucial to get people from different backgrounds talking – and listening – to each other. If the chatter is anything to go by, the conference is a roaring success. People are swapping notes on their posters, figuring out where their research fits together and asking keenly after each other's breakthroughs. Everyone keeps talking about the prevalence of PhD students and early-career researchers. These young chemists don't have the same confidence as their supervisors, but they're enthusiastic and clear: however their careers unfold, they know sustainability will dictate them.

At the conference dinner, an impromptu dance party starts. The senior academics lead it off, but – once enough of them have been coaxed to join – the younger crowd dances the longest.

Big-picture thinking

Prior to green chemistry, chemists were not monolithically careless about the environment. There were always concerned people who tried to minimise the effects of what they were doing. Professor

Qin Li, an environmental engineer at Griffith University who is co-chairing the conference with Raston, knows this well: her late father was a chemistry professor who worked with paints.

'He was always telling me, "Oh, this is toxic, don't touch that." When we had fruits and vegetables, I was always made aware of the pesticides on the surface,' she says. 'Green chemistry is something in my veins.'

Li's own research on particle movement started to get seriously environmental after her PhD, when she was looking at groundwater and aquifer recharge, and how her particle technology could improve clogging. She's carried this interest in water through her career – I catch her just after a breakfast workshop, where she and a dozen others have been discussing ways to repurpose wastewater.

'Where I think it is a paradigm shift is in asking the people who come up with the recipe to think in a holistic way, in a multi-dimensional sphere,' she says. 'They really need to look at: okay, it's not only just about removing dirt, but also about after removing the dirt. What happens to that surfactant? Because if it's going to end up in the environment, then it will also come back to us.'

A future of solutions

The hotel tells me they have no use for my umbrella, so I stuff it into a coat pocket and take it on the flight home. On the plane, I draw up a plan for it based on things I've heard or seen at the conference.

The aluminium stretchers are simple to deal with – aluminium is almost as easy to recycle as it is to mine, and much less carbon-intensive. They'll be separated out and melted down, as will the steel in the springs. The fabric, probably polyester, can be burned into hard carbon, which could become the anode for a sodium-ion battery. This process still releases some CO_2, but that could be captured and added to Air Company's vodka. The hard plastic is trickier. I could use some sort of catalyst that breaks the polymers down (preferably not a heavy metal) – once broken, they could become carbon-based catalysts themselves.

I'm being flippant, designing this plan for 200 grams of plastic when I'm on a machine emitting several kilograms of CO_2 each second. But fixing the problem of jet fuel won't fix the problem of my junk umbrella, either. The path to a sustainable future is paved with several thousand small solutions.

They might be alchemists, but chemists have also always been crafters. They make molecules. For decades, these molecules have had one job: to do something as efficiently as possible. Now, more and more chemists are starting to think about everything else that happens to their molecules. Do they break down in the environment? Is there waste involved in making them? What else might they affect?

It's in the interests of both chemists and the companies they work for to start crafting molecules in this paradigm. Otherwise, they're betting against the rain: packing their hyper-efficient materials into a suitcase and thinking, foolishly, that the climate won't interfere. If more and more chemists are making things sustainably, the old methods will become, well, unsustainable.

'We just laugh, or cry, when people say, "Oh, chemistry is a mature field,"' says Anastas. 'It is so nascent. It is at the beginning of its potential.'

✱   *Born to ruler?*, p. **229**
    *Chips now come in flavours like cheeseburger. How do food
      chemists get the taste right?*, p. **256**

INDIGEMOJI

India Shackleford

Kaytetye speakers, artists, technologists and linguists have teamed up to create the Kaytetyemoji app! They hope to teach a new generation of Kaytetye language speakers through the art of emoji.

Endangered languages

Australia is home to more than 250 Indigenous languages, and 800 different dialects. Each language is specific to its People and place.

Language is an important part of identity. For thousands of years, cultures all around the world have used language to pass on stories, knowledge, lore and geography.

Unfortunately, a lot of these languages are in danger.

Kaytetye (kay-ditch) is one such endangered language, with only 109 speakers recorded in the 2021 census. It's spoken by the Kaytetye People, whose Country is north of Mparntwe/Alice Springs in Central Australia.

It would be easy to see phones and computers as a threat. But Kaytetye speakers reckon that technology could be the key to saving their language instead!

Making an emoji

The first emoji was created in 1999 by Japanese artist Shigetaka Kurita. Initially, emoji were limited to simple icons representing things such as weather, traffic and the time. Since the 1990s, the emoji list has continued to grow, with new additions like the goose and moose emoji in 2023.

It's a lengthy and difficult process to get new emoji added to the Unicode official list. For this reason, the Kaytetyemoji team decided to make their own sticker set based on Kaytetye language and culture. At the beginning of 2022, a group of Kaytetye speakers met and began to brainstorm ideas for Kaytetyemoji.

They contacted Indigemoji, who'd previously created Australia's first Aboriginal emoji set for the Arrernte (ah-ruhn-duh) language. Together, they worked to design, code and translate emoji to match important words in Kaytetye life and culture.

Kaytetyemoji

The Kaytetyemoji app contains 112 different emoji, each representing different words and aspects of life on Kaytetye Country. The app allows users to see and hear words spoken in the language and includes an example of the word in a sentence.

'One of the benefits of the app is having short and simple things to learn,' says Dr Myfany Turpin, who is a linguist (a scientist who studies language).

'It's an entry point if you're starting from scratch, but you can also go further and listen to sentences!'

A favourite emoji of the Kaytetyemoji team members is artnke, which means flat-topped hill. Artnke are a key feature in the landscape around Barrow Creek.

'You'll see them while you are driving,' Myfany explains.

They are symbols of Kaytetye Country, in a similar way that the Opera House is a symbol of Sydney.

Connecting with kids

Phillip Janima is a Kaytetye speaker who provided audio for the emoji, along with his father and niece. He hopes the app will encourage young people to engage with language and culture.

'We're trying to help all these young ones going to school [which is in English] to learn [Kaytetye],' says Phillip.

He also emphasises the connection between culture and language.

'[When] people do things that involve Kaytetye language, it's perfect for it – like ceremony and hunting. When those activities stop ... part of the language stops.'

GIF away

The team is also keen to explore GIFs as another way of digitising the Kaytetye language. Hand signs are an important part of Kaytetye language and can help listeners to tell the difference between homophones. These are words that sound the same, but mean different things, such as horse and hoarse, or knew and new!

✱   *Predicting the future*, p. **122**
A chemist's guide to optimism, p. **261**

CONTRIBUTORS

MATTHEW WARD AGIUS is a journalist from Adelaide. He has been a staff writer for *Cosmos Magazine*, and has had work appear in Australian Community Media mastheads, the *Advertiser*, *InDaily* and *The Wire Current Affairs*. He currently works at Deutsche Welle in Germany.

JACINTA BOWLER has spent almost a decade as a science journalist and has written about everything from AI in art to zero emissions technologies. They have worked for the ABC, *Cosmos Magazine*, *Nature*, ScienceAlert and many more. They were also published in *The Best Australian Science Writing 2023*.

JOSEPH BROOKES is an Australian journalist covering technology, innovation and research. He is currently a senior reporter at InnovationAus.com where he covers technology and innovation policy, including science and research funding. His writing focuses on government policy, procurement, digital inclusion and research funding, with a focus on the impact of government decisions. In 2020 Joseph was recognised as the best new information technology journalist and has since been runner-up in the technology issues and news coverage awards.

TABITHA CARVAN is senior science writer for the Australian National University and a freelance writer on the side. Her work has been featured in publications including the *Guardian*, the *Sydney Morning Herald* and the *Age*, *The Saturday Paper*, *Crikey*, *Junkee*, *Australian Geographic* and *The Best Australian Science Writing 2022* and *2023*. She is also the author of the memoir *This is not a book about Benedict Cumberbatch* (Fourth Estate).

CARLY CASSELLA is an independent science journalist who tells stories about the natural world and our place within it. She has written for *bioGraphic*, *High Country News*, ScienceAlert, *Australian Geographic*, and more. Born in the Pacific Northwest and raised in Australia's sleepy bush capital, Carly is rarely in one place for long. Her home for now is Naarm/Melbourne Australia on the traditional lands of the Kulin nation.

LAUREN CHAPLIN is the Senior Communications Coordinator at the Garvan Institute. A former tech journalist, her work has featured across more than fourteen print, digital and e-commerce titles, including news.com.au, *The Daily Telegraph*, the *Herald Sun*, Finder and Women Love Tech.

VIKI CRAMER is a writer and ecologist whose work seeks to understand how both the human and more-than-human world can flourish in the landscapes we share. In 2021 Viki was awarded a Dahl Fellowship from Eucalypt Australia. Her first book, *The Memory of Trees: The future of eucalypts and our home among them*, was published by Thames & Hudson Australia in 2023 and shortlisted for Book of the Year in the WA Premier's Book Awards 2024.

ANGUS DALTON is a science reporter at the *Sydney Morning Herald* who has ventured deep into sewers, flooded swamps and broiling climate chambers in search of good stories. He was the cofounding editor of *Sweaty City*, a magazine about urban ecology and climate change, and his work has appeared in *Australian Geographic*, *Overland* and *Kill Your Darlings*.

KATE EVANS is a science and nature writer based in New Zealand with both Kiwi and Australian heritage. She has written for the *New York Times*, *Scientific American* and the *Guardian*, among many other outlets, and has just published her first book, on feijoas.

ELIZABETH FINKEL is a biochemist who switched to journalism. She co-founded *Cosmos Magazine*, serving as editor-in-chief from 2013 to 2018. She authored *Stem Cells*, which won the Queensland Premier's Literary Award, and *The Genome Generation*. In 2019 she received the Medal of the Australian Society for Medical Research and an honorary doctorate from Monash University. She serves as vice-chancellor's fellow at La Trobe University and on an advisory committee for Latrobe University Press.

PROFESSOR JENNY GRAVES PhD, FAA, AC compares genes and chromosomes of distantly related (especially Australian) animals to discover how sex works and how it evolved, (in)famously predicting the loss of the human Y chromosome. Among many honours and awards, she won the 2017 Prime Minister's Prize for Science. Author of 465 research publications, Jenny also expresses her love of beautiful science in many popular articles and the occasional poem and libretto.

LYDIA HALES is a freelance journalist based in lutruwita/ Tasmania. She has worked as an in-house reporter for ABC News, *Australian Doctor* and the *Medical Republic*, and her freelance writing has appeared in *The Best Australian Science Writing*, the *Guardian*, *Australian Geographic* and *Cosmos Magazine*. She also once had a poem published on a tree.

RICH HARIDY is a freelance writer and journalist based in Melbourne, Australia. His work focuses on science, technology and the new world of psychedelic science. He currently sits on the Science Journalists Association of Australia committee and his writing can be found at places including *Nature*, *Salon*, *Cosmos Magazine*, and *New Atlas*. His website is richharidy.com.

AMALYAH HART is a freelance science writer based in Melbourne. Her work runs the gamut from energy and climate to consciousness and AI, and has been featured in the *Saturday Paper*, *Cosmos Magazine* and others.

JUSTINE E HAUSHEER is an award-winning science writer for The Nature Conservancy. Her favourite stories take her into the field: following logging elephants through Myanmar, surveying for sea cucumbers in Manus, and wading into outback waterholes. She holds a masters degree from New York University and a bachelor's degree from Princeton University. Justine is writing a book on threatened species conservation, to be published in 2026 by NewSouth.

LEIGH HAY is a freelance writer and editor. The author of nine books, she is a published author and poet. A former May Gibbs Writer-in-Residence, Leigh and husband Dr David Hay host the Tales from the Treehouse website <www.talesfromthetreehouse.com.au> – a writers' cooperative to promote the works of Australian authors and poets. An experienced chorister, Leigh wrote the libretto for *When the Bugle Calls,* premiered in 2016, to commemorate the battles of the Somme and Long Tan.

ZOE KEAN is a science writer and communicator based in lutruwita with a special passion for evolution, the environment and ecology. Her debut book *Why are we like this?* will be published in late 2024.

ANGE LAVOIPIERRE is the ABC's national technology reporter, and an award-winning comedian and writer (*Gruen*, *The Monthly*, the *Guardian*). Ange's AI reporting for Background Briefing won an Australian Podcast Award in 2023. She's the creator and host of the ABC's *Schmeitgeist* podcast, about trends and phenomena in tech and internet culture, and attracted a Walkley nomination for their investigation into ADHD care. Previously, Ange was the host of the ABC's daily news podcast, *The Signal*.

MICHAEL LEACH has a PhD in pharmacy and is a senior lecturer at Monash University's School of Rural Health. Michael is also an award-winning poet. Michael's poems reside in journals such as *Cordite Poetry Review*, anthologies such as *Poetry for the Planet* (Litoria Press), and his books *Chronicity* (MPU), *Natural Philosophies* (RWP), and *Rural Ecologies* (ICOE Press). Michael's next poetry book is forthcoming from Ginninderra Press. He lives in Bendigo on Dja Dja Wurrung Country.

DYANI LEWIS is an award-winning science journalist based in Melbourne. She is a regular contributor to *Nature*, where she reports on evolution, the environment, health and matters of science that impact public policy. Her work has also been published by the *Atlantic* (via *Undark*), *Science*, *Cosmos Magazine*, the *Guardian*, *The Monthly*, *Smith Journal* and others. Her first book, *Unvaxxed: Trust, truth and the rise of vaccine outrage* (Hardie Grant), was published in June 2022.

LIAM MANNIX is the national science writer for the *Age* and the *Sydney Morning Herald*. His science reporting has won multiple prizes, including the Eureka Prize for Science, the Walkley for short feature writing and the Quill for science journalism. He usually lives in Melbourne but is in Denmark on a fellowship with the Constructive Institute, looking at ways to improve coverage of the climate crisis. His first book, *Back Up*, on the science of back pain, was published in 2023.

SHEY MARQUE is an award-winning poet from Western Australia. *The Hum Hearers*, her second poetry collection, was shortlisted for the Dorothy Hewett Award 2023 and is forthcoming with UWA Publishing in 2025. *Keeper of the Ritual* (UWA Publishing, 2019), was shortlisted for the Noel Rowe Poetry Award. She worked as a scientist in clinical and research facilities within Australia and New Zealand. She holds a BAppSc (Hons), PhD Molecular Pathology, and an MA Writing.

AMANDA NIEHAUS is a writer, editor and co-publisher of the magazine *Science Write Now*, which publishes creative writing inspired by science. Her acclaimed first novel, *The Breeding Season* (Allen & Unwin, 2019) is based on the unusual reproductive lives of northern quolls. Amanda lives in Brisbane/Meanjin.

BIANCA NOGRADY is a freelance science journalist and author whose work has appeared in publications including the *Guardian*, the *Saturday Paper*, the *Atlantic*, *Nature*, *WIRED* and *MIT Technology Review*. Her writing has been selected for nine editions of the *Best Australian Science Writing* anthology, and she has edited the anthology twice. Bianca was founding president of the Science Journalists Association of Australia, and she wants climate action now.

NATALIE PARLETTA completed a PhD in 2005 and after ten years of her own research branched out to freelance writing with the help of a Post-Baccalaureate Certificate in Science Writing with Johns Hopkins. She has written for various publications, including *Cosmos Magazine*, the *Guardian*, the *New Daily*, *Health Agenda*, the *Sydney Morning Herald*, *The Age*, *Ensia* and *Nature*, and won the Australasian Medical Writer Association's 2019 Early Career Award and the Crawford Fund's Food Security Journalism Award.

ELLEN PHIDDIAN is a science communicator who practises her craft with science writing, shows, events, podcasts and videos. Holding degrees in chemistry and science communication, she worked in outreach for various organisations before joining the *Cosmos Magazine* newsroom for a three-year stint as a journalist. She currently lives in Adelaide.

JAMES PURTILL is the ABC's online technology reporter, covering stories from social media to solar panels, and artificial intelligence to electric vehicles. In 2023 he hosted an ABC podcast on the rise of AI, 'Hello AI Overlords', and travelled to Papua New Guinea to write and record three long-form science stories. He's worked as a reporter in Port Hedland, Kalgoorlie, Darwin and Sydney, and now lives in Perth, Western Australia.

DREW ROOKE is a journalist, author, and *The Conversation*'s Assistant Science & Technology Editor. His books include *One Last Spin: The power and peril of the pokies* (Scribe, 2018) and *A Witness of Fact: The peculiar case of chief forensic pathologist Colin Manock* (Scribe, 2022), which was shortlisted for the 2022 Ned Kelly Award for Best True Crime. He was a 2021 Our Watch Fellow.

INDIA SHACKLEFORD is a medical student, health researcher and freelance science writer. As a Kaurna/Ngarrindjeri writer, she is passionate about showcasing the work and knowledge of Aboriginal and Torres Strait Islander STEM professionals past, present and emerging.

BELINDA SMITH became a science journalist after realising she wasn't going to cut it as a scientist. She's currently a journalist with ABC Science, and her work appears on the ABC News website. You can also hear her talking about science on local radio and RN. In her spare time, Bel's a GPS artist who runs maps in the shape of animals. Find them on Insta @animalpunruns.

ALICIA SOMETIMES is a poet, artist and broadcaster. She has performed her poetry at venues, festivals and events around the world. Her poems have been in *Best Australian Science Writing*, *Best Australian Poems* and many more. Alicia is director and co-writer of the art–science planetarium shows *Elemental* and *Particle/Wave*. In 2023, she received ANAT's Synapse Artist Residency and co-created an art installation for Science Gallery Melbourne's exhibition *Dark Matters*. Her new book is *Stellar Atmospheres*.

CAMERON STEWART is the chief international correspondent and also a senior feature writer with the *Australian* and the *Weekend Australian Magazine*. He has held many roles with the newspaper, including as a two-time US correspondent, based in New York and then in Washington. He is a former winner of the Graham Perkin Award for Australian Journalist of the Year.

PETRA STOCK is a climate and environment reporter at the *Guardian*. Previously at *Cosmos Magazine*, her work has also featured in *Australian Geographic*, *The Age*, *Crikey*, the *Mandarin* and *Archer*, among others. Before absconding to journalism, Petra had a 20-year professional career working in environmental approvals, Aboriginal heritage policy, renewable energy and climate change. She is also a qualified environmental engineer.

CLARE WATSON is a freelance science journalist with a background in biomedical science. Since trading her pipettes for a pen, her work has aired on ABC Radio National's *Health Report* and appeared in *TIME* (via Undark), *Nature*, the *Guardian*, *Hakai Magazine*, *Cosmos Magazine* and *Australian Geographic*. She writes and fact-checks for ScienceAlert, and loves finding stories that pull back the curtain on how science happens and who scientists really are.

ACKNOWLEDGMENTS

We'd like to thank the wonderful team at UNSW and NewSouth for their ongoing support of the *Best Australian Science Writing* and Bragg Prize for Science Writing. They are vital celebrations of this craft. We both remember the first time we saw our names as contributors in a book and we're thrilled NewSouth continues to make that moment possible for future generations of science writers.

Our thanks must also go to every contributor, and the publishers who provide a platform and space for science writing, many of whom are listed below.

Finally, our utmost appreciation to the Science Journalists Association of Australia and the vibrant community that has grown around science writing in Australia. Since its inception in 2019, it has supported and championed the work of science journalists across the country and provided opportunities for young writers to learn and grow. This is crucial for a healthy and sustainable science writing culture. May it long continue.

'In the heart of the forest, one woman built a house of slime' by Liam Mannix was originally published under the same title for *The Age* and the *Sydney Morning Herald* on 13 November 2023 and is available at <www.smh.com.au/national/in-the-heart-of-the-forest-one-woman-built-a-house-of-slime-20231102-p5egzx.html>.

'Satellite tracking the Pacific's most endangered leatherback turtles' by Justine E Hausheer was originally published under the same title for *Cool Green Science* on 16 April 2023 and is available at <blog.nature.org/2023/04/16/satellite-tracking-the-pacifics-most-endangered-leatherback-turtles/>.

'Western Australia had its hottest summer ever, but climate change barely made the news' by James Purtill was originally published under the same title for ABC News on 18 March 2024 and is available at <www.abc.net.au/news/science/2024-03-18/wa-summer-heat-broke-records-but-media-downplayed-climate-change/103572922>

'The heroes of Zero' by Cameron Stewart was originally published under the same title in the *Weekend Australian Magazine* on 5 August 2023.

'Dog people: How our pets remind us who we really are' by Amanda Niehaus was originally published under the same title in the *Griffith Review* on 7 November 2023.

'Can a wild animal make your house feel like a home?' by Tabitha Carvan was originally published under the same title in *ANU Science* on 25 March 2023 and is available at <science.anu.edu.au/news-events/news/can-wild-animal-make-your-house-feel-home>.

'Of moths and marsupials' by Kate Evans was originally published under the same title at *BioGeographic* on 26 April 2023 and is available at <www.biographic.com/of-moths-and-marsupials/>.

'The consciousness question in the age of AI' by Amalyah Hart was originally published under the same title for *Cosmos Magazine* on 2 October 2023.

'AI's dark in-joke' by Ange Lavoipierre was originally published under the same title at ABC News on 15 July 2023 and is available at <www.abc.net.au/news/2023-07-15/whats-your-pdoom-ai-researchers-worry-catastrophe/102591340>.

'Why solar challenges? They're in the DNA of Tesla, Google for starters' by Matthew Ward Agius was originally published under the same title for *Cosmos Magazine* on 23 October 2023 and is available at <cosmosmagazine.com/technology/why-solar-challenges-theyre-in-the-dna-of-tesla-google-for-starters/>.

'Poetic constellations – exploded sonnet sequence' by Shey Marque was originally published under the same title for *Westerly Magazine* on 1 November 2023.

'Origins – of the universe, of life, of species, of humanity' by Jenny Graves and Leigh Hay was originally published under the same title for Melbourne Recital Centre on 18 July 2023. The full live version is available online at <www.youtube.com/watch?v=YRH7MD7F4U0>.

'Rural pharmacy placement' by Michael Leach was originally published under the same title for *Partyline* on 17 March 2023.

'The world's oldest story is flaking away. Can scientists protect it?' by Dyani Lewis was originally published under the same title for *Nature* on 6 December 2023 and is available at <www.nature.com/immersive/d41586-023-03818-5/index.html>.

'Predicting the future' by Drew Rooke was originally published under the same title for *Cosmos Magazine* on 14 March 2023.

'This little theory went to market' by Elizabeth Finkel was originally published under the same title for *The Monthly* on 1 November 2023.

'How scientists solved the 80-year-old mystery of a flesh-eating ulcer' by Angus Dalton was originally published under the same title for the *Sydney Morning Herald* on 30 January 2024 and is available at <www.smh.com.au/national/how-scientists-solved-the-80-year-old-mystery-of-a-flesh-eating-ulcer-20240129-p5f0qr.html>.

'"Why would you find me attractive?": The body disorder that needs more attention' by Lydia Hales was originally published under the same title for the *Guardian* on 29 October 2023 and is available at <www.theguardian.com/australia-news/2023/oct/29/why-would-you-find-me-attractive-the-body-disorder-that-needs-more-attention>.

'Why my adenomyosis went undetected for five years' by Lauren Chaplin was originally published under the same title for News.com.au on 8 March 2024 and is available at <www.news.com.au/lifestyle/health/health-problems/why-my-adenomyosis-went-undetected-for-five-years/news-story/ca238feb2a74b095073857e6353eacf1>.

'Doing drugs differently: For public health, not profit' by Clare Watson was originally published under the same title for *Cosmos Magazine* on 1 December 2023.

'Indigenous science must be a standalone national science priority' by Joseph Brookes was originally published under the same title for *InnovationAus* on 11 October 2023 and is available at <www.theguardian.com/australia-news/2023/jul/23/bush-learning-boom-forest-school-kindy-beach-outdoor-immersion>.

'Are psychedelics a treatment for long COVID? Researchers probing this mystery don't have the answer yet' by Rich Haridy was originally published under the same title for *Salon* on 7 December 2023 and is available at <www.salon.com/2023/12/07/are-psychedelics-a-treatment-for-long-covid-researchers-probing-this-mystery-dont-have-answers-yet/>.

'Cecilia Payne-Gaposchkin' by Alicia Sometimes was originally published under the same title in *Stellar Atmospheres* (Cordite Press) on 1 March 2024.

'The last King Island emu died a stranger in a foreign land' by Zoe Kean was originally published under the same title for ABC Radio Hobart and ABC News on 5 June 2023 and is available at <www.abc.net.au/news/2023-06-05/the-king-island-emu-died-alone-in-paris/102425608>.

'Call of the liar' by Carly Cassella was originally published under the same title for *BioGraphic* on 2 August 2023.

'Science in the balance' by Jacinta Bowler was originally published under the same title for *Cosmos Magazine* on 11 March 2024.

'Everything starts with the seed' by Natalie Parletta was originally published under the same title for *Cosmos Magazine* on 14 March 2024.

'Born to ruler?' by Petra Stock was originally published under the same title for *Cosmos Magazine* on 14 December 2023.

'Surviving in one place' by Viki Cramer was originally published in her book *The Memory of Trees* on 27 June 2023.

'"Give the espresso a little swirl": The very particular science of a good cup of coffee' by Bianca Nogrady was originally published under the same title for the *Guardian* on 8 January 2024 and is available at <www.theguardian.com/science/2024/jan/08/how-to-make-a-good-cup-of-coffee-espresso-steps-details-tips>.

'Chips now come in flavours like cheeseburger. How do food chemists get the taste right?' by Belinda Smith was originally published under the same title for ABC News on 15 December 2023 and is available at <www.abc.net.au/news/science/2023-12-15/crisps-flavour-chemistry-aroma-taste-smell-laboratory-csiro/103031004>.

'A chemist's guide to optimism' by Ellen Phiddian was originally published under the same title for *Cosmos Magazine* on 14 September 2023.

'Indigemoji' by India Shackleford was originally published under the same title for *Double Helix* magazine (*doublehelix.csiro.au*) on 1 September 2023.

2024 Shortlist

In 2012, UNSW Press launched an annual prize for the best short non-fiction piece on science written for a general audience. The UNSW Press Bragg Prize for Science Writing is named in honour of Australia's first Nobel laureates, William Henry Bragg and his son William Lawrence Bragg. The Braggs won the 1915 Nobel Prize for physics for their work on the analysis of crystal structure by means of X-rays. Both scientists led enormously productive lives and left a lasting legacy. William Henry Bragg was a firm believer in making science popular among young people, and his lectures for students were described as models of clarity and intellectual excitement.

The UNSW Press Bragg Prize is supported by the Copyright Agency Cultural Fund and UNSW Science. The winner receives a prize of $7000 and two runners-up each receive a prize of $1500.

The shortlisted entries for the 2024 prize are included in this anthology.

The UNSW Press Bragg Prize
for Science Writing 2024 Shortlist

Kate Evans, 'Of moths and marsupials'
Dyani Lewis, 'The world's oldest story is flaking away.
Can scientists protect it?'
Liam Mannix, 'In the heart of the forest,
one woman built a house of slime'
Amanda Niehaus, 'Dog people'
Drew Rooke, 'Predicting the future'
Cameron Stewart, 'Heroes of Zero'

Winners announced in November 2024.
<unsw.press/the-unsw-press-bragg-prize-for-science-writing/>

Judges for the UNSW Press Bragg Prize 2024

Professor Merlin Crossley, University of New South Wales
Professor Joan Leach, Australian National University
Distinguished Professor Lidia Morawska, Queensland University
of Technology
Jackson Ryan, co-editor, *The Best Australian Science Writing 2024*
Carl Smith, co-editor, *The Best Australian Science Writing 2024*

The successful Bragg Prize, which recognises excellence in science communication, hosts a special category for high school students. Science enthusiasts in years 7 to 10 are invited to submit an essay of up to 800 words.

A joint initiative of UNSW Press, UNSW Science and Refraction Media, with support from the Copyright Agency Cultural Fund, the prize is designed to encourage and celebrate the next generation of science writers, researchers and leaders. For an aspiring university dean of science or Walkley Award–winning journalist, this could be the first entry on their CV.

The winner receives a $500 UNSW Bookshop voucher, publication in next year's *The Best Australian Science Writing* and CSIRO's *Double Helix* magazine, and an invitation to the launch of *The Best Australian Science Writing* in Sydney. Two runners-up each receive a book voucher worth $250. The regional and city schools with the most entries receive a book pack from NewSouth Publishing, including a copy of this book. The names of the winners and details about the competition are available on the UNSW Press website: <unsw.press/the-unsw-bragg-student-prize-for-science-writing/>.

The 2023 competition invited students to write a short essay on 'How does science and technology use or benefit from AI and how should we best navigate a future where AI is part of our everyday lives?' Entries in 2023 were judged by a panel comprising Donna Buckley, mathematics and cybersecurity teacher at John Curtin College of the Arts; Sarah Chapman, co-chair of Women in STEM; Brooke Jamieson, senior developer advocate, Amazon Web Services; Donna Lu, editor of *The Best Australian Science*

Writing 2023; and Heather Catchpole, Royal Society shortlisted author and founder of Careers with STEM.

The winning entry was 'Two paths to AI: The choice of humanity' by Elsie Paton, a year 9 student from Kambala School, Rose Bay, NSW. Her winning essay is included on the following pages.

TWO PATHS TO AI: THE CHOICE OF HUMANITY

Elsie Paton
Year 9, Kambala School, Rose Bay, NSW

Conjure a mountain slope. On one side, a snow-covered path descends to a city where AI is harnessed to aid the sick, make scientific discoveries, provide unbiased feedback, protect the community, and improve the environment. As you shift your gaze over to the other side of the mountain, you glimpse another city where people use AI for different purposes: surveillance, the storage of sensitive information, personal gain and the production of prejudiced or apathetic decisions. You stand on the top of the mountain, gazing at the two cities, with your own feet leading the way

Well, firstly, what is AI? Artificial intelligence is information-processing technology that uses algorithms to perform cognitive tasks such as prediction, decision making and data analysis. Though often assumed to be unethical, artificial intelligence has aided science through predicting interactions between chemicals, performing experiments, and diagnosing conditions with greater precision, efficiency, superior speed and reduced financial expenditure, therefore providing countless ethical benefits to many communities around the world.

For example, in pathology, AI software was used by Insilico Medicine to design a drug that slows the development of idiopathic pulmonary fibrosis, a chronic disease that affects the lung's alveoli and can lead to respiratory failure and death. Furthermore, AI4Good's 'Climate Trend Scanner' aids global warming research, whilst Deepmind's investigation into proteins that speed up the development of plastic-degrading enzymes helps sustainably combat environment-related issues. Evidently, a plethora of possibilities awaits as AI reveals its potential to attract multiple ethical benefits to society, through improving science, our actions, lives and surroundings.

However, despite AI's obvious ethical help in many sectors of science, there are multiple concerns regarding AI's potential bias, lack of emotion, misuse, job replacements and, lastly, the ethics of 'human enhancement'.

Firstly, many AI models have been found to repeatedly internalise and express existing prejudices in society. For instance, Georgia Technology discovered that self-driving cars had a higher rate of collision with people of a darker skin tone due to a poorly created dataset, and in 2019, an AI algorithm was found to favour white patients over black patients when advising whether follow-up care was necessary! Simply picture the consequences of AI's prejudiced decisions in fields of science, medicine and day-to-day life: could they eventually impact you?

Secondly, although artificially intelligent technologies currently lack emotion, their input is being taken to heart in delicate situations. For instance, picture a world where an AI 'death algorithm' predicts the likelihood of patient survival and thus gives input on whether life support should continue. Its programming lacks empathy, hope and understanding, yet the doctors trust its accuracy. Well, imagining this should be straightforward, as it already exists, and is strongly opposed. This is because not only is AI unable to recognise the true impact of its actions, but its possible prejudices could further lead to potentially poor decisions in such important situations. Some have questioned whether AI's role in life-or-death situations is even ethical at all, due to the possibility of severe damage.

Thirdly, there has been speculation as to whether AI could replace jobs due to its superior performance. However, the ethics of replacing those who need a source of income with machines that don't could be questioned.

Fourthly, AI poses privacy issues due to its potential for misuse and violation. As AI systems contain vast amounts of sensitive data, personal information such as a patient's date of birth could be hacked, or important surgeries and scientific experiments could be manipulated and spied upon through technology. Algorithms and scientific equipment could be maliciously programmed for personal gain, resulting in devastating, immoral impacts for science and greater society.

Lastly, and perhaps most importantly, the ethics of 'human enhancement' through AI must be considered whilst neuroscience delves deeper into technologies that alter the brain. AI and precise gene editing are two very recent discoveries in science with ethical implications. The recent creation of a gene editing program named 'CRISPR' allows diseases such as Huntington's and HIV to be permanently eliminated from embryos. However, according to STAT News, 2020, researchers at Columbia University found that 'in more than half of the cases, the editing caused unintended changes, such as loss of an entire chromosome', leading to harm. Consequently, it must be considered whether it is fundamentally ethical to modify our being, and whether this could create a disparity between those who are 'enhanced' and those who are not.

AI poses similar ethical challenges. Whilst AI is capable of, and is, bringing significant scientific advancement to the world, aiding countless lives, and nurturing our environment, there are a multitude of palpable possibilities in which AI can be unethically abused. Our continuing embrace of AI calls for caution, yet it simultaneously reveals a kaleidoscope of colours, possibilities and benefits, where the world could be helped in ways unbeknownst. Ultimately, it is up to us, humans, to select the mountain path which lies before us, and cultivate AI's great potential in an ethical way.

ADVISORY PANEL

MERLIN CROSSLEY AM is Deputy Vice-Chancellor Academic Quality at UNSW. He studies CRISPR gene editing and blood diseases. Crossley is an enthusiastic science communicator, Chair of *The Conversation*'s Editorial Board and Deputy Director of the Australian Science Media Centre. He was a Rhodes Scholar, a recipient of the NSW Premier's Award for Medical Biological Science, and in 2021 a species of iridescent squid was named in his honour – *Iridoteuthis merlini* – Merlin's butterfly bobtail squid.

JOAN LEACH is Director of the Australian National Centre for the Public Awareness of Science at the ANU. Professor Leach has been president of Australian Science Communicators and chair of the National Committee for History and Philosophy of Science at the Australian Academy of Science. Her research centres on public engagement with science, medicine and technology, and she has published extensively about science communication, including the recent volume *An Ethics of Science Communication* (with Fabien Medvecky).

LIDIA MORAWSKA, a physicist, is a Distinguished Professor and Australian Laureate Fellow at the Queensland University of Technology and the director of the International Laboratory for Air Quality and Health at QUT, a WHO Collaborating Centre. She conducts fundamental and applied research in the field of air quality and its impact on human health and the environment, and is a member of the Australian Academy of Science and the American Academy of Arts and Sciences.